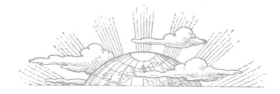

SHORES OF KNOWLEDGE

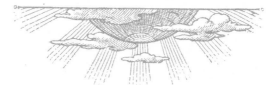

SHORES OF KNOWLEDGE

New World Discoveries

and the Scientific Imagination

JOYCE APPLEBY

W. W. NORTON & COMPANY

NEW YORK · LONDON

For information about permission to reproduce selections from this book,
write to Permissions, W. W. Norton & Company, Inc.,
500 Fifth Avenue, New York, N Y 10110

For information about special discounts for bulk purchases, please contact
W. W. Norton Special Sales at specialsales@wwnorton.com or 800-233-4830

Manufacturing by RR Donnelley, Harrisonburg, VA
Book design by Mary Austin Speaker
Production manager: Anna Oler

Library of Congress Cataloging-in-Publication Data

Appleby, Joyce Oldham.
Shores of knowledge : new world discoveries and the scientific imagination /
Joyce Appleby.
pages cm.
Includes bibliographical references and index.
ISBN 978-0-393-23951-5 (hardcover)
1. America—Discovery and exploration—European. 2. Discoveries in
geography—History. 3. Discoveries in science—History. I. Title.
E101.A66 2013
970.01—dc23

2013017212

W. W. Norton & Company, Inc.
500 Fifth Avenue, New York, N.Y. 10110
www.wwnorton.com

W. W. Norton & Company Ltd.
Castle House, 75/76 Wells Street, London W1T 3QT

1 2 3 4 5 6 7 8 9 0

I dedicate this book to my son,
Mark Lansburgh, in gratitude
for his many, precious gifts.

Contents

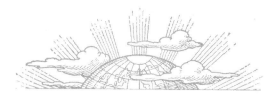

SHORES OF KNOWLEDGE

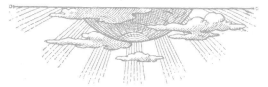

Introduction

When Christopher Columbus returned from the Western Hemisphere in the spring of 1493, he came with news that would decisively change Europe. No consequence would be more portentous than the conversation his discoveries prompted about the natural world, for he made the subject of nature suddenly interesting with all the odd things he brought home. Sailing back to Spain on the *Niña*, he packed the little caravel to the gunwales with fantastic objects from the Caribbean islands he visited. Six Taino natives, out of a dozen, survived the return trip, giving vivid proof that people lived in what geographers called the antipodes. Birds from the West Indies survived the trip better than the Tainos. Columbus had plucked flowers even more colorful than the brilliant parrots he found in the tropical rain forests. He showed his sponsors, King Ferdinand and Queen Isabella, a bit of gold that some natives had given him, thinking that it would guarantee funding for subsequent voyages—and he was right.

Over the course of the next three centuries, a succession of amateur investigators laid the foundation for the modern life sciences even though before the end of the fifteenth century, Europeans had been an incurious people. Finding these masses of land filled with mysterious people, unfamiliar plants, weird animals, and striking topography produced the kind of shock essential to shaking free of the church's venerable injunction against asking questions about nature. Men and women in China and Muslim Spain's Córdoba had demonstrated a much stronger inquisitive spirit. In Europe, isolation and religious disapproval had curtailed curiosity for over a millennium.

Today, plying incessant questions to nature is one of the stron-

gest features of the modern West. The stirring of Europeans'
interest in the physical world began at the end of the fifteenth
century with the discovery of two continents lying between them
and the Orient. Like most profound cultural changes, there were
layers of habits and convictions to work through before Europeans could engage fully with the natural world. They had to break
with the church's prohibition of intrusive questioning about
God's domain—the phenomena of his created universe. The
assumption that they already knew everything worth knowing
had erected another barrier to the investigative spirit, as well as
the predisposition to look backward to biblical or classical times
for guidance and knowledge. The designation New World suggests the dimension of their surprise. That the forebears of those
in the West, long distinguished for its scientific élan, had to be
blindsided before they became inquisitive comes as a surprise. It's
also true, and this is a book about how it happened.

The Catholic Church had succeeded for a thousand years in
keeping curiosity in check out of fear of probing questions about
cosmic events like eclipses and comets. Such inquiries were
deemed vain, a petty challenge to God's all-encompassing knowledge. Of course the church couldn't suppress all curiosity, certainly not a child's endless queries about what and why. But even
the spirits of children will be dulled if the answers they hear are
always "because God willed it so." The campaign against curiosity
began with Augustine, who lived when Christianity was becoming the dominant religion in Europe.

In its heyday a century before Augustine was born, the Roman
Empire stretched from Britain to North Africa and eastward from
the Iberian Peninsula to both the Black and Red Seas. Germanic
tribes began to make incursions into Rome's domain in the third
century. Emperor Constantine established a new capital on the
Bosporus. Then the Empire split permanently into eastern and
western parts, with the eastern center of religious life in Constantinople. Constantine, whose mother converted him, legalized

Christian worship in 313. As the Eastern Orthodox Church grew in importance, the Western Empire declined. By the time Augustine, born in North Africa, had become a passionate Christian, the Visigoths had sacked Rome.

The accelerating decline of the Roman Empire had a profound effect. The humbling of this great power suggested the ephemeral quality of human endeavors. The replacement of the civility of Roman rule with the utter confusion wrought by the victorious Germanic tribes turned men and women's thoughts to God's order. Augustine evoked the imagery of the City of God and the City of Man to console his anxious contemporaries. He comforted them by emphasizing the transience of earthly things and hence their unworthiness of a true Christian's concern.

More relevant to this study, Augustine, now a bishop, condemned curiosity about the material world. He stated unequivocally that "it is not necessary to probe into the nature of things, as was done by the Greeks." "Nor need we be in alarm," he elaborated, "lest the Christian should be ignorant of the force and

Conrad Hietling's engraving of "Pilgrim flanked
by Devotion and Curiosity," 1711.

number of the elements—the motion and order and eclipses of
the heavenly bodies; the form of the heavens, the species and the
nature of animals, plants, stones, fountains, rivers, mountains,
about chronology and distances, the signs of coming storms; and
a thousand others things which those philosophers either have
found out or think they have found out." "It is enough," he concluded, "for the Christian to believe that the only cause of all created things ... whether heavenly or earthly ... is the goodness of
the Creator, the one true God."[1]

At another time such a dogmatic position might not have rever-
berated through the ages, but the centuries after the fall of Rome
were chaotic. The invading Germanic tribes, coming in waves over
three centuries, had to build permanent settlements far from their
origins, while Romans and their allies dealt with the destruction
of their civilization. The church offered a haven from the turmoil. The bar to curiosity entrenched itself as an article of faith.
In the ensuing centuries, Europeans pretty much stuck to home,
concentrating their intellectual energy on establishing Christian
institutions that would withstand doubt, despair, and dissent.

Isidore of Seville advised leaving "to one side, like a secret,
anything which the authority of the Holy Scriptures has not
caused you to learn." He called curiosity a dangerous presumption
leading to heresy. "It embroils the mind," he said, "in sacrilegious
fables."[2] Isidore lived in the seventh century, so his strictures
might be of little import for people eight centuries later. But that
in fact was the problem: Christians were supposed to adhere to a
set of beliefs frozen in time.

Bernard of Clairvaux evoked the fall of Adam to castigate
curiosity as "the beginning of all sin." God had punished Adam
and Eve severely for it. Men intent on learning the "height of the
sky, the breadth of the earth, and the depth of the sea" earned the
disapproval of the thirteenth-century pope Innocent III as well.
Monks were given manuals to help them stifle curiosity, which
was thought to arouse inappropriate desires, both intellectual and

sensual. Christians might not know everything about the world, but their God did, and that should be sufficient. Everything that existed was subordinate to God's will; anything that happened had providential implications.[3]

A very powerful and pervasive institution, the church claimed the authority to discriminate between legitimate and illicit knowledge, between permitted and prohibited questions, even between accepted and forbidden methods of acquiring knowledge. After the Reformation, Protestant leaders revived the attack on curiosity in the sixteenth century. John Calvin associated it with the deadliest of deadly sins: pride. King James I of England pointed to Eve for evidence of how curiosity could harm someone.

Europeans were little exposed to the larger world through travel. Religion had stirred crusades to the Holy Lands in the tenth, eleventh, and twelfth centuries, but after that, the crusaders' descendants stayed put.[4] Overland trade with Asia was cut off for decades at a time. Still, there were some venturous souls. Merchants in Genoa and Majorca began visiting the islands off Africa. The Vivaldi brothers of Genoa set sail west to India, but were never heard from again.

When that most celebrated traveler, Marco Polo, returned to Europe in 1294, he landed in a Genoese prison. Fortunately for thousands of future readers, his cell mate happened to be a writer. To him, Polo recounted the details of his Venetian merchant family's twenty years in the Orient and how his father met Kublai Khan, the grandson of Genghis Kahn. He described their encounters as diplomats and traders at the great Mongol court at Karakorum and the high drama of their escape from their possessive host in a perilous two-year voyage. Most Europeans got their first impression of China, India, and Japan from this travel journal.[5]

Henry the Navigator, a Portuguese prince, spent his considerable fortune sending expeditions down the west coast of Africa in the middle decades of the fifteenth century. He was determined to find out what lay south of Cape Bojador at the 24th parallel, which marked the farthest that Europeans had sailed down the west coast of Africa, but he dispatched others there, preferring to stay at home himself.

Henry was fired by the desire to wrest the Canary Islands from Spain and to find a coastal source for the gold and slaves traded in the interior of the African continent. Where he differed from his predecessors was in recognizing the importance of improving navigation. While fighting for Portugal against the Arabs in Morocco, he took stock of what their mariners knew about commercial linkages and the level of their sailing skills. He familiarized himself with Arab mapmaking. Never going to sea himself, Henry was content with gathering around him in his academy at Sagres a cadre of expert navigators, shipwrights, astronomers, pilots, and cartographers, both Christian and Jewish. He was mindful of fears that the southern waters were filled with monsters and wrapped in deadly fog, but he calmly stated, "You cannot find a peril so great that the hope of reward will not be greater."[6] Like Columbus a half century later, ambition fueled his many navigational projects.

The ships that plied the Mediterranean were far too slow and large for oceanic travel. Venturing out in this unknown stretch of the Atlantic demanded new techniques and new equipment, accrued through trial and error. Arab navigators in the southern Mediterranean had introduced triangular sails, fore and aft, called lateen sails. They also added a small foremast, the mizzen, that improved steering.[7] Henry incorporated these Arab inventions into his light, fast caravel, rigged for sailing close to the wind. From his estate perched on the rocky promontory jutting into the Atlantic in the southwest corner of Portugal, he sent pairs of these caravels to chart the winds and waters along the bulge of Africa.

From the 1420s until his death in 1460, Henry's expeditions got ever larger; they located successful routes and found safe harbors for provisioning. Slowly, after many failed endeavors, they solved the problems of navigation in the South Atlantic. Once they brought back gold and slaves, the voyages became remunerative.[8] Within a century, Africans composed a tenth of the population of Lisbon, a city still underpopulated from the Black Death of the previous century.

Henry died before Bartolomeu Dias rounded the Cape of Good Hope in 1487, but by that time explorations down the coast of Africa had found other royal patrons. When the Portuguese mariners got below the equator, the North Star was no longer in the heavens, which meant that they had to develop new celestial navigation with the Southern Cross, the constellation visible in the Southern Hemisphere almost any time of the year.

Adding to the pressure to find a sea route to the Indies was the fall in 1453 of Constantinople to the Ottoman Turks, who closed off trade with Europe. Bringing in cinnamon, cloves, nutmeg, and pepper on overland routes became more costly and more fraught with peril.[9] Going by water to the fabulous islands could cut out the Arab merchants who acted as middlemen. Having discovered the winds that would carry ships around the tip of the continent, Dias opened the way for others to reach the East Indies. A generation of sailors from the ports of Italy and the Iberian Peninsula could now follow where the geographic pioneering of Henry had pointed.

With prospects for great success, the Portuguese king had no compunctions about rejecting appeals for support from a Genoan named Christopher Columbus, who had a different idea about reaching the Indies. Discouraged, Columbus turned to Spain, where Queen Isabella and King Ferdinand had just joined their kingdoms to form a united monarchy in Spain.[10] As rivals of Portugal, they sought a different route to the fabled riches of the Orient and offered to sponsor Columbus's expedition. Even more important to the devout queen than catching up with the Portuguese was the possibility of extending the realm of Christendom in Asia. Columbus shared this goal. He extracted promises from them to receive 10 percent of all the goods found in the lands.

His son wryly noted years later when the family was fighting over what they thought was due their father that the monarchs had probably not expected him to get back alive. But he did, after a seven-month round trip to the Caribbean.

News of the discoveries in the Western Hemisphere arrived when Europeans were still absorbing the philosophy of ancient Greece and Rome. After being cut off from these stores of knowledge and wisdom for a millennium, scholars acquired access to them through the libraries of the Muslims in Córdoba. The rebirth in the term "Renaissance" refers to the flowering of art and literature in response to this recovery of classical texts. After the Turks took Constantinople, many Greek scholars moved to Italy, bringing with them a thorough understanding of the Greco-Roman writers whose worldly perspective was so different from the spiritual otherworldliness of medieval Europe.

Christians might not know everything about the world, but their God did, and that should be sufficient. Everything that existed was subordinate to God's will; anything that happened had providential implications. This presented difficulties to Thomas Aquinas and others living in the thirteenth century who were trying to incorporate the newly accessed texts from Aristotle into the Christian tradition, for Aristotle had considered the desire to know a natural one. Aquinas's solution was to distinguish among possible objects of curiosity on the basis of their contribution to Godliness.[11]

Ancient writings brought in a questioning attitude that startled with its intellectual insouciance. They bristled with inquiries, hypotheses, and stratagems for proof. To read them was to reassess what one thought one knew. The freshness of this engagement has to be measured against the medieval obsession with doctrine, form, and faithfulness to sacred texts. An appreciation of the Greek and Roman philosophers, like news of the voyages of discovery, moved from the fifteenth-century preserve of classicists to a wider group of educated readers with the publication of translations of the texts.

As secular concerns intruded upon religious ones, a naturalism came to dominate paintings and sculpture. Michelangelo's famous statue of David comes to mind. Philosophical prose became more direct and vivid. The desire to make the world a better place in which to live led some to ponder which forms of government were the best, an analysis that tended to undermine the awe that all governments had deliberately cultivated, the better to rule.

Greater knowledge of the ancients led to the recognition of how profoundly different in their tastes, mores, assumptions, and convictions they were from fifteenth-century Europeans. The divergence of Greek learning from Christian cosmology added to the intellectual turmoil from the discovery of the New World. Slowly—social and intellectual change is always slow—it dawned upon Europeans that their world was a far different one in every measure conceivable. From this conclusion came the even more arresting awareness of change over time, of history itself, not as a narrative of events, but as a depiction of a bygone era that reveals the various ways that different societies have shaped human experience. The past became like a foreign country, but it took a much more thorough examination of it to turn this insight into a meaningful metaphor.

Europeans began studying the recovered works of Ptolemy, the first-century Greek who had collected in one volume all the geographic knowledge that had been acquired in his Greco-Roman world, just as the discovery of the New World was making more salient than ever how little they actually knew about the planet they lived on.[12] With an old text and a new map, they began delving into just those subjects that Augustine had excoriated. Even Spanish churchmen who followed in the wake of the explorers became curious. They justified this new intellectual trait on the grounds that none of their authorities—biblical or ancient—knew anything about these strange continents accidentally discovered on the way to the Orient.

An avid appreciation for ancient philosophers created a whole new category—that of the humanists, who cultivated classical ideas and styles through a rigorous study of Greek and Latin. Greek and Roman thinkers had been famously curious, asking a myriad of questions about the heavens, the planets, and earth's human inhabitants. Reading their works couldn't but nurture an investigative spirit. Responding to this, the humanists formed clubs to talk about the knowledge gleaned from the recovered texts. Their growing numbers signaled a discontent with the inward, logical reasoning of the Scholastics, who were the principal interpreters of the Christian dogma that had dominated thinking since Augustine's time.

The new focus on the classics was not like sailing off into the Western Hemisphere, but studying Greek philosophy prepared some to ponder the puzzles that the discoveries turned up. Especially startling was Europe's new location on the globe, no longer joined to the Middle East and Asia through a vast land bridge, but separated from Asia by two huge oceans and the linked continents of North and South America. Maps had to be redrawn and redrawn again as successive explorers returned with new sightings. Ironically, the fact that the existence of the New World was unknown to the ancient writers dented their reputation a bit. Greek geographers had planted serious doubts about there being life near the equator. Desiderius Erasmus, the great Dutch humanist, wondered in what other ways the ancients might be in error.[13] This attitude of questioning, if pressed too far, led to conflicts with the Scholastics.

Turning toward this world, as ancient learning encouraged the humanists to do, meant attending to the objects people encountered every day—the animals, rocks, mountains, trees, and stars—not to mention fellow human beings. Leonardo da Vinci's many anatomical drawings exemplified this new fascination with the here and now. His accurate depictions of the human body merged art into scientific inquiry. Not until the nineteenth century did philosophers, mathematicians, artists, and scientists go their separate ways. For three centuries, gifted amateurs in all these fields took the lead in examining natural objects, both domestic and foreign.

Printing with movable type, introduced in the second half of the fifteenth century, made the reproduction of writings and illustrations much less expensive than the written manuscripts that had preserved texts before. During these same decades, the reading public expanded with the switch from Latin to vernacular languages. Publishing and literacy enhanced one another as the catch basin for communication widened. The divergence of Greek learning from Christian cosmology added to the intellectual turmoil from the discovery of the New World. More and more men and women had to cope with the intrusion of novelty, but it would be a mistake to exaggerate the immediate impact. Still, there was a momentum going for new initiatives in exploration.

Greeks had long pointed to the Pillars of Hercules, the promontories flanking the Mediterranean's opening to the ocean, with the warning ne plus ultra—go no farther. Portugal and Spain, facing the Atlantic, became the obvious kingdoms to reject this advice. During the 1580s and '90s the Portuguese had been building victualing stations on African islands and for their commercial fleets en route to the trading centers in the East Indies. Meanwhile, a French navigator, Jean de Bethencourt, sailing for Castile in 1402, conquered the Canaries, which lay some 1,200 miles from Cadiz, winning for the Spanish a key station in the Atlantic.[14] From the Canaries Spanish sailors could catch the best winds to carry them west. By the end of the fifteenth century, monarchs and financiers were ready to open up their purses to expeditions that would explore the world by sea.

Like Prince Henry, Columbus studied navigation. He made maps to depict his conjectures of what the globe really looked like. He also got in touch with Paolo de Pozzo Toscanelli, a Florentine astronomer and geographer of note, who had come into contact with writings long lost to Western Europe. Lorenzo the Magnificent had summoned an ecumenical council in 1439. The thirty-one Greek bishops who attended this extraordinary gathering brought with them the knowledge of ancient philosophy that had been preserved in the Byzantine Empire. They knew the speculations and experiments of the inquisitive Greeks. Toscanelli talked to these bishops in Florence, reigniting his zeal to figure out the shape of the earth.

It was one thing to know that the world was not flat and another to have an accurate idea of its size and shape. Europeans had neither. Worse, their heads were full of hideous pictures of what lay beyond the waters that lapped at their shores. Only with the great persuasive powers of a prince of the realm had Henry the Navigator got his sailors to press farther down the west coast of Africa. Most people believed that fantastic creatures inhabited the Ocean Sea, as they called the Atlantic. Others were sure that the bottom teemed with sinners who were burning in a molten mass that could suck in vessels that sailed out too far. Access to the writings of Strabo, a first-century Greek, dissuaded Toscanelli of the existence of these terrors. Strabo insisted that there was one world and it was habitable and its landmasses were joined by the Ocean Sea. Toscanelli went further and said that it would be possible to sail from Europe to India along the same parallel.[15]

Men with grand visions like those of Columbus had existed before; he succeeded in implementing his plans because financiers, merchants, and monarchs—usually given to caution—responded positively to his outsized ambition. Columbus's plan contained two errors: he calculated that Japan was 2,400 nautical miles from the Canaries when it was actually over 10,000. As problematically, he did not anticipate there being a landmass between Europe and Asia! With the confidence of ignorance, he departed in August 1492 with some 120 men dispersed among two little caravels, the *Niña* and the *Pinta*, and the larger *Santa María*.

Ardor for heroic adventures may have helped suppress the fear of Columbus's seamen for what lay beyond the coastal waters. His sailors were lucky that Columbus knew about the clockwise circular wind patterns of the Atlantic, which would get them back home before exhausting their food supply. Elated by his success in making a landfall after nine weeks, Columbus sailed home convinced that he had found a landmass not far away from Asia whose store of riches he had read about in his favorite text, *The Travels of Marco Polo.*

In 1497 the king of Portugal commissioned Pedro Álvares Cabral to strengthen contacts with Asian merchants, sending him off with a fleet of thirteen ships. Blown off course in a storm, Cabral landed in Brazil and forthwith claimed it for Portugal before pushing on to India. Eager to tamp down the already disruptive competition between the two Catholic countries of the Iberian Peninsula, Pope Alexander VI had divided the globe between Portugal and Spain. Now he had to make a large jag in the established line (longitudes were then guessed at) to honor Portugal's new possession.

Columbus's discoveries were extraordinary enough to batter at the wall of inhibitions that surrounded questioning of Christian cosmology. There was no place in the European system of knowledge to fit in the plants, animals, minerals, and humans he brought back. They challenged settled opinions and provoked unbidden questions; they tugged at the roots of faith. The voyages of discovery proved to be the catalysts for breaching the church's curbs on curiosity, but it took time.

The intellectual consequences of Spain's venture across the Atlantic long outlasted its empire. Slowly the age-old concern with acquiring wisdom through contemplation was pushed aside in the pell-mell search for mundane details about the earth and its contents. Old ways of knowing were turned upside down. A passion for collecting information through observation, measurement, and description of new phenomena grew stronger, though it took successive generations to generate hypotheses about their meaning. This new form of pursuing information opened up the doors of inquiry to less educated amateurs, once excluded from the closed circles of the Scholastics and humanists who had to master ancient languages.

The engagement with natural phenomena involved Europeans in an inquiry about sex and sexuality. The nude bodies of the Amerindians provoked questions about the meaning of nakedness. Innocence and barbarity competed as answers. When explorers encountered willing sexual partners in the New World, their reports led to a new discourse about sexuality. Nor were sexual questions confined to humans. Botanists would use reproductive organs to categorize plants, and the sexual exclusivity among animals became the way naturalists defined species. Reproduction and the incorporation of new traits in the lineage of living things would form the basis of Darwin's explosive explanation of human origins, bringing to a climax the four-century examination of natural phenomena that the discoveries of the world outside of Europe provoked.

No one had any idea of what would happen if ships sailed west across the Atlantic. Columbus's sponsors certainly did not expect the most significant unintended consequences of all: the breaking open of the closed world of Christianity. A civilization marked by a reverence for sacred texts so deep that it disallowed questions about natural phenomena became the trailblazer in inquiries about nature. The Church Fathers had been correct. Curiosity was dangerous. Passing from amateur passions to sober investigations of biology, geology, and astronomy, it upended the grand Christian narrative of the origins of life and the place of our planet in the universe. Over the course of four hundred years the research spawned by Columbus's discovery of the New World set Europe apart from any other society on the globe, and, even more, from its own past.

CHAPTER ONE

THE NEW WORLD FINDS

ITS SCRIBES

Columbus in India primo appellens, magnis excipitur muneribus ab Incolis. IX.

RIMA nauigatione, quum Columbus terram attigit, crucem ligneam in littore statuit: deinde prouectus in Hoytin Insulam appellit, quam Hispaniolam nuncupat, & in terram cum multis Hispanis descendit. Ibi quum ab eius loci Cacico (regulum ita appellant) cui nomen Guacanarillo, summa comitate exceptus esset, muneribus inuicem datis & acceptis, ambo sidem amicitiæ futuræ sanxere. Columbus, indusiis, pileolis, cultellis, speculis & similibus eum donauit: Cacicus contra satis magno auri pondere Columbum remuneratus est.

C 2   Colum-

Depiction of Christopher Columbus greeting the Taino Indians in
Hispaniola, taken from The Grandes Voyages [1590–1598]
published by the Belgian engraver Theodor de Bry.

Once news of Columbus's voyage passed through the ports and courts of Southern Europe, rough-cut seamen, adventurous merchants, penurious hidalgos, and scions of noble families tumbled out of the Iberian Peninsula like oranges from an overturned basket. Some were soldiers no longer needed after the successful expulsion of the Moors from Granada in 1492. Others went in search of land and people to work it. Thousands of Spanish with a sprinkling of Italians, Germans, French, and Portuguese began plying the waters between the Atlantic coast of Europe and the Gulf of Mexico. To English speakers their names are verbal bonbons: Diego de Almagro, Juan Ponce de León, Gonzalo de Ocampo, Francisco Hernández de Córdoba, Hernán Cortés, Amerigo Vespucci, Luis Vásquez de Ayllón, Alvar Nuñez Cabeza de Vaca, Vicente Yañez Pinzón, Bernal Díaz del Castillo, Francisco Peñalosa, Ruy López de Villalobos, and Vasco Núñez de Balboa.

While the conquistadors were busy turning the Caribbean into a Spanish lake, intrepid merchants ventured out to size up the commercial possibilities now that their rivals, the Portuguese, had found an all-water route to the Orient. The church soon sent priests and missionaries. Those with the most enduring influence were the men who produced eyewitness accounts of the New World novelties and a few writers whose scholarly bent prompted them to write histories of the Europeans' new world.

Within six months of his landfall in the Caribbean, Columbus set sail for Asia again, this time with seventeen ships and twelve hundred soldiers, two hundred of them gentleman volunteers. His royal sponsors, reluctant to fund his initial voyage, now enthusiastically backed Spanish settlements. Some volunteers joined Columbus's second expedition to start the settlement that King Ferdinand and Queen Isabella were eager to establish in their newly claimed lands. From such a beachhead they could bring the Amerindians into the Christian fold while enriching their depleted treasury with gold and other precious goods. Columbus packed the seventeen ships with seeds, plants, horses, dogs, pigs, cattle, chicken, sheep, and goats. (Even while nurturing his precarious settlement, Columbus found time to continue his search for Japan and China on two more voyages across the Atlantic.)

The many ships crossing the Atlantic for commerce, conquest, or piracy after 1492 breached forever the isolation of the Amerindians, while their existence compelled thousands of Europeans to come to their habitats to exploit them. Printers immediately produced engravings to satisfy the curiosity of Europe's reading public. At first they trotted out old lithographs that had been used to depict the Garden of Eden. Gradually, more accurate depictions of the people, animals, and plants from the Western Hemisphere were published, adding some specificity to the rumors, myths, and tall tales that circulated almost immediately.

These trips that carried people, plants, animals, and—more menacingly—germs from the Old to the New World and back again started what has aptly been called the Columbian exchange. The centuries-old mingling of Asians, Africans, and Europeans had produced a kind of immunity to each other's diseases. No such protection existed for the Tainos, who were the first to feel the deadly force of European germs. Other exchanges were more benign.

As his cargos suggest, Columbus was the original Johnny Appleseed, bringing familiar plants and animals to sustain the Spanish in their new settlements while carrying back home chilies, beans, potatoes, tomatoes, corn, and tobacco, which then made their way to Asia and Africa. With few natural predators in Hispaniola, European animals luxuriated on the native grasses and roots. Even the stowaway rats did well. If it is true that an army marches on its belly, the rapidly multiplying swine kept the Spaniards' stomachs full. Horses would become critical to the their successful conquests when they moved out from Hispaniola to Cuba, Panama, Venezuela, and the splendid kingdoms of Mexico and Peru.[1]

The Caribbean's peculiar people, unfamiliar plants, bizarre animals, and striking topography produced enough shocks to overcome the prevailing predisposition to accept the world as a given whose essential features could be gleaned from biblical or classical texts. Christians had incorporated Asians and Africans into their cosmology, but they had no place in their system of knowledge for the puzzling items that navigators delighted in bringing back from the New World.

Initially the Tainos, Caribs, and Carawaks of the West Indies had seemed like good candidates for the lost tribes of Israel. One explorer described a group of Indians with faces "gentle and noble as those of classical sculptures." Their nakedness took on special meaning as a sign of natural virtue. Their handsome bodies announced that they were good candidates for conversion to Christianity.[2] Eventually the Spanish degraded them to children of the devil whose indolence and ignorance justified their enslavement. And then there was the distinctive vegetation and animals not known to the Old World, troubling to Christians whose Bible was presumed to describe all that humans needed to know.

Two boys—eight-year-old Bartolomé de Las Casas and fifteen-year-old Gonzalo Fernández de Oviedo—witnessed the grand reception accorded Columbus in 1493 when he returned to claim the title Admiral of the Ocean Sea. Oviedo was at court the day Columbus presented the six Tainos to the king and queen and laid before them gold trinkets, chilies, and masks made of precious stones and fish bones. At fifteen, Oviedo already fancied himself the historian of the Spanish empire in the making and began taking notes about Columbus's voyage and the Indians, parrots, masks, and gold jewelry that attested to the success of the voyage to parts unknown. Las Casas, for his part, remembered that when Columbus came to Barcelona, the Spanish monarchs honored him by offering him a seat next to them when he gave his report.[3]

Las Casas and Oviedo probably didn't know each other then, but they were destined to meet as adversaries in the Spanish Indies, where each would find his life's work. Most of their contemporaries sailed west to start trades, lay out farms, or fill the ranks of the empire's hierarchies of church and state. Oviedo and Las Casas did some of these things, but they also wrote. Their reports—and especially their histories—carried the fascinating details about the wondrous New World to a European reading public eager for authoritative voices to inform them.

When Oviedo witnessed Columbus's triumphal return, he had already created an important place for himself at the Spanish court as a courtier for Don Juan, the son and heir of King Ferdinand and Queen Isabella. Oviedo expected that he would remain attached to the prince and move up at court as Juan matured and eventually became king. Alas, the prince died unexpectedly in 1497, when Oviedo was nineteen. Thrown out of his cushy job, Oviedo bemoaned his bad luck and felt compelled to leave his country and wander about looking for new sponsors. Bright and ambitious, he set off for Italy, where the grass was considerably greener than on the Iberian Peninsula.[4]

Worse things could befall a young man than living in Italy in the closing years of the fifteenth century, a span of time so glorious that Italians gave it a special name, the *Quattrocento*. Oviedo soaked up the humanist culture brightened by luminaries like da Vinci and Michelangelo, whom he met while working in various royal households in Italy. Intellectually curious, he could not but profit from these encounters with men who were changing the way Europeans looked at the world. In da Vinci, Oviedo saw a man so imaginative, so talented, so learned that his achievements as an artist, engineer, inventor, architect, musician, writer, sculptor, and student of nature gave rise to the concept of the Renaissance man.

The sight of Columbus's return when he was fifteen had stirred Oviedo's imagination, prompting him to correspond with Vicente Yañez Pinzón, one of the three captains in Columbus's first fleet. When Oviedo returned to Spain, King Ferdinand asked him to write a history of the Castilian monarchy, which he did.[5] He married, but the death of his wife left him a widower at twenty-seven. Then King Ferdinand sent him to the Caribbean as a notary and overseer of the gold foundries. Oviedo arrived in the New World in 1514 in an expedition of twenty-two ships carrying two thousand Spaniards—many of them future conquistadors. The hope of finding a fortune enticed many an hidalgo to seek permission to conquer for the king. They found disease and death more often than pots of gold, but they planted the king's colors and a cross wherever they went—with little cost to the monarchy.

A whole new universe unrolled before Oviedo when he landed in Panama. Though he sailed to the Caribbean as a royal official and remained one until his death in 1557, curiosity overtook his life once he stepped off the boat. Panama, then called Darién, formed part of what the Spanish named Terra Firme, an area extending from the Yucatán Peninsula to Venezuela. From his first weeks Oviedo felt compelled to keep a record of the novelties that assailed him at every turn. He had long been an accomplished writer; now he had his topic: the Spanish in the New World. His stated goal was to help his king understand what he now possessed and to keep alive the memories of the great Spanish achievements in the New World.[6] Oviedo didn't go to university. His Latin— still the mark of an educated person—was sketchy at best. Instead he wrote in Castilian Spanish, which ended up giving him a larger audience of vernacular readers.

It didn't take Oviedo long to jump into the ferocious infighting among the Spanish officers in the Indies. He had left Spain with the new governor, Pedrarias Dávila, whom he quickly judged to be cruel and corrupt. This conviction sent him back to Spain twice to plead for Dávila's dismissal before Spain's new monarch, Charles I, who became Duke of Burgundy, Lord of the Netherlands, ruler of the rapidly expanding territories in the Americas, as well as Holy Roman Emperor, ruling as Charles V. These titles came to him from his mother, Joanna, the daughter of Ferdinand and Isabella, and his father, Philip the Handsome of Burgundy. Attending court now meant traipsing all over Europe as Charles moved from one possession to another.

While in Spain, Oviedo married again. Success crowned his effort to get Dávila dismissed, but the appointed successor died within a few month of his arrival. Death hovered over the entire Spanish enterprise. Oviedo had taken his second wife and their two sons back to Darién with him in 1520. There he set up his household in a fine mansion. Within two years all three had died, victims of a smallpox epidemic. Oviedo was disconsolate. Only an eleven-year-old illegitimate son survived. Mortality rates for Europeans were not nearly as high as were those for the Indians, but they could still be explosive in some places. Oviedo estimated that of the 2,500 original settlers at Darién, only eighteen survived.[7] Hundreds more Spaniards were killed in futile attempts to find fortunes in places like Florida and Bolivia, where the indigenous people were well prepared to defend their lands against the invaders.

Europeans expired by the thousands, but Native Americans were dying by the millions. European diseases such as measles, typhus, and smallpox proved to be lethal killers. The newcomers in the Western Hemisphere had triggered an unintended holocaust, for they carried with them microorganisms against which the natives had no protection. Conquistadors carried their deadly force of European germs to the Aztecs, Mayans, and Incas. Even men and women in those spots the Spaniards left unoccupied, like the islands of the Lesser Antilles, were felled by disease when slave-hunting bands transmitted their European germs to them.

To decimate a population is to kill one tenth of it. In the New World the ratio was reversed: 90 percent of some Indian tribes died; others were entirely wiped out. From an estimated high of 40 to 112 million people in pre-Columbian America, the population fell to 4.5 to 10 million people a century later.[8] The fatal encounter between Europeans and the indigenous population of North and South America repeated itself again and again well into the nineteenth century, when Russians arrived in Alaska. Without an understanding of the invisible organisms that carry disease, neither the Spanish nor the native people understood the disaster that was unfolding around them.

Spain's seaborne adventures drew on Europe's venerable aristocratic traditions. The noble virtues of command, mastery, ardor, and aggression had marked the long Spanish effort to expel the Moors from their country. They were also qualities that proved very effective in intimidating and subduing people who were similarly impressed by manly feats of valor. The *Reconquista* of the Iberian Peninsula was in many ways a rehearsal of the conquests of Mexico and Peru. Columbus was typical in never running out of new places to conquer. He was also a religious visionary. He calculated that the potential riches in the New World would pay for an expedition of one hundred thousand crusaders to capture Jerusalem for Christendom.[9]

In the ten years between his first and fourth voyages, Columbus turned to making a fortune. Eager to exploit the gold mines that he found on Hispaniola, he started capturing Indians to work for their new Spanish masters. He initiated a system that would enrich, enrage, and haunt the men who built the Spanish Empire. With the king's confirmation, the *encomienda*, a regime of virtual slavery, became the basis of the Spanish economy in the New World whether the work at hand was mining, raising cattle, or tending commercial crops. The Spaniards destroyed Indians' homes to force them to live in settlements that were accessible for labor and for religious conversion.

As the natives continued to die in great numbers, slave-hunting expeditions fanned out to all the islands of the Caribbean. Columbus even began sending enslaved Tainos to Spain, but Queen Isabella became queasy about this trade and returned many of them.[10] Quickly exhausting the gold deposits on Hispaniola, the charter settlers looked for a surer source of profits. Columbus pioneered the production of sugar with cane that he carried from Portuguese Madeira. More than any other crop, sugar sustained the profitability of the Spanish Empire, ensuring as well the endurance of the *encomienda* and the African slave system that succeeded it.

Oviedo arrived in Hispaniola with the unenviable task of announcing to Native Americans their new status as subjects of the Spanish king. As he described it in his *General and Natural History of the Indies*, he read the document aloud to deserted towns. A farcical ritual repeated over and over, the proclamation-reading usually acted as the prelude to warfare. Oviedo's job also included branding newly captured Indians. As the people became more knowledgeable of Spanish ways, they fought back, leading to savage campaigns of retribution by the Spanish. Oviedo was a reluctant witness to these atrocities, but he wholeheartedly believed in the Spanish mission of settling the New World and bringing Christianity to its people.

Though neither soldier nor sailor, Oviedo shared the lingering chivalric spirit. Offered the governorship of Colombia, he petitioned the emperor for a hundred knights to subdue the Indians and form the political and social structure for the colony. When Charles declined to supply the knights, Oviedo turned down the appointment. Invited to take up the post again three years later, he posed the even more bizarre condition of banning from the region all university men, attorneys, and friars. Nothing came of this, so he contented himself with writing a novel about knights and a treatise on heraldry.

Most Spaniards in the New World shared Oviedo's complicity in subduing the indigenous peoples. Las Casas was the exception. His inability to turn a blind eye to the suffering the Spanish wrought upon them changed him from a prosperous settler into a passionate crusader. Las Casas's father had gone out to the New World on Columbus's second voyage, returning with an enslaved Taino boy for his son, who was a student at Salamanca. Juanito became one of those natives whom King Ferdinand insisted be returned. He was sent back, but not before young Bartolomé had become familiar with his ways.

Coming out on Columbus's fourth and last voyage, Las Casas arrived at Hispaniola during the first raw decade of Spanish "pacification," when entire groups of native people virtually disappeared. They mainly died from European diseases, but Spanish violence and unremitting forced labor competed for their lives. Disease was a silent killer; abuse a more conspicuous one. Seeing their children among the murdered added to the despair of Indian men and women, many of whom committed suicide rather than endure enslavement.

Already a deeply religious man, Las Casas turned to the church for a career after he had been in the New World for eight years. In fact he was the first man from the New World to seek ordination as a Catholic priest. The ideas of the twenty-six-year-old about the native people took shape over the next decade of observation and reflection. Like most of his contemporaries, he was not opposed to slavery per se. While he was growing up, there

were close to ten thousand African slaves in Seville, woven into
the everyday experience of families like his. Las Casas inherited
land from his father and worked it with the slaves he had been
granted when he came to Hispaniola. Had the Spanish treated
their slaves with care and kindness, he might have spent his clerical career tending a flock of Spaniards and their households. His
acceptance of the status quo collapsed when he participated as a
chaplain in the conquest of Cuba.

After two decades of rapacious development, the Spanish in
Hispaniola began to feel the loss of the native workers who had
succumbed to "pacification," disease, and harsh working conditions. By 1508, Spaniards had circumnavigated Cuba and realized
that it was a big island, not part of the continent to the north,
as Columbus had believed. Rumors that Cuba had gold mines
induced Diego Velázquez de Cuéllar with a small force to invade
it in 1511. Tainos who had fled the abuse on Hispaniola were prepared to resist, unlike their peaceful relatives who had greeted
Columbus nineteen years earlier.

The campaign was brutal. In the village of Caonao, while van-
quished natives unsuspectingly brought food to the invaders, the
Spanish outdid themselves in gratuitous slaughter. Arms, legs,
shoulders, breasts were hacked off men and women; three thousand were left to die. The Spanish intended to send a message to
the Tainos still fighting. "I saw here cruelty on a scale no living
being has ever seen or expects to see," Las Casas wrote. The massacre at Caonao haunted him. He clung to the horror of watching
a young Taino whose right shoulder had been cut off. For years he
tormented himself because he had left with the Spanish forces,
offering no help to the wounded man.[11]

Finally the Spanish captured the leader of the Taino resistance,
a chieftain named Hatuey, who was condemned to be burned at
the stake. A Franciscan friar told him that he would go to heaven,
and not hell, if he became a Christian. Hatuey asked if it was

true that Christians went to heaven. Hearing that they did, he declined the offer saying that he would prefer to go to hell so he would never see such a cruel people again.[12]

With Cuba conquered, the Spanish proceeded to organize the surviving natives into search parties to locate gold in the streams of the central highland ranges. Spaniards swarmed into Cuba, almost emptying Hispaniola. A new pattern emerged. Within a few years, the conquistadors had skimmed the surface wealth in Cuba and were ready to move on to Mexico when they got news of Cortés's conquest in 1521.[13]

A Spanish official worrying about the depletion of Indian laborers got authorization to import more slaves from Africa. The king established an annual contract open to slave-trading countries to bring four thousand slaves into the Spanish Caribbean. Spain's first *asiento*, or contract, went to Portugal; later England won the lucrative contract. One of the first to take advantage of this royal permission, Vasco Núñez de Balboa used thirty African slaves to build ships on the Pacific. Thus began a trade that would bring eleven million Africans to the New World.

By 1565 there were seven times more Africans than Europeans in the New World. Las Casas initially supported the introduction of an African labor force because he thought that blacks were stronger and more accustomed to hard work than the natives of the Caribbean. As he later saw Africans being worked to death in the Americas, he realized that their suffering was no different from that of the Indians.[14]

After the Cuban invasion Las Casas began grappling with the evils of Spanish rule. The pope had granted the Spanish crown control of the Catholic Church in the New World. The imperative to bring Christianity to the pagans brought in question the efficacy of the *encomienda* system. For his participation in the Cuban invasion, Las Casas, an *encomendero* himself, got more land and labor allotments. This prompted a Dominican missionary to refuse him the sacraments for holding slaves. Angered at first, Las Casas came to accept the Dominicans' principled opposition to the exploitation of the natives. Struggling with his newfound sense of guilt, he realized that he would have to give up his slaves if he were to join the Dominicans' crusade.

Las Casas's decision appalled Governor Velázquez de Cuéllar, who told him that he was on his way to becoming a rich man and reminded him that, as a "secular priest"—in contrast to members of a monastic order—serving Spaniards, he didn't have to act like the Dominican missionaries pledged to shun the world's evils. What Las Casas was doing, Velázquez said, was a "thing new and, as it were, monstrous." Las Casas remained firm. He shared his newfound clarity about the urgency of reform with his parish-ioners, whom he admonished to confront "the cruelty they were committing against those innocent, meek people."[15] When they, *encomenderos* all, ignored calls to give up their slaves, Las Casas left the Indies to speak directly to the king. And thus at thirty he began his long crusade.

The return to Spain in 1515 had many consequences. It was during this five-year visit that he and Oviedo clashed publicly before King Charles in Barcelona. The issue undergirding their dispute was whether Native Americans had the capacity to lead decent, independent lives. Las Casas maintained that they did, but Oviedo pointed to their nakedness, idolatry, and sloth to argue the opposition. This confrontation triggered an animosity that lasted until death. Las Casas spent more time opposing Oviedo than any other adversary, in part because Oviedo spoke with the same authority of personal experience that Las Casas drew upon.

While Las Casas was in Spain, Hernán Cortés published the first of the letters that would make him famous. A fine writer, Cortés joined the cluster of eyewitness chroniclers from the New World with his stirring rendering of the conquest of Mexico that involved a succession of arduous battles, complicated by duplicity, intrigues, conspiracies, massacres, sieges, uprisings, and narrow escapes. He evoked the Spaniards' marvel when they finally occupied the capital, Tenochtitlán, and saw the handsome buildings, extensive markets, resplendent residential sections, and "floating gardens" that produced the city's food. The Aztecs' ceremonial human sacrifices repelled the Spanish, but the massive stone statues, gold artifacts, and handsome frescoes impressed them.

Cortés looked upon the Mesoamericans he had conquered as his personal possessions. Reputed to have the largest holding of indentured Indians in the New World, he owned three thousand outright, with possibly as many as twenty-four thousand more forced to leave their villages for a week at a time in order to work on his estate. There they cut wood, built structures, planted sugarcane, and milled it for him.

Las Casas, Cortés, and Oviedo capture in their writings the complex of motives and goals of the sixteenth-century Spanish leadership. The king sought glory, gold, and a clean conscience. Cortés fought for fame, money, and power. Oviedo wanted to prosper and ascend the ladder of the official hierarchy, but writing a history of the glorious deeds of men like Columbus and Cortés became an end in itself for him. Las Casas became a writer in order to pursue justice for the native people whose way of life he saw disintegrating before him. Both Las Casas and Oviedo were drawn, as well, to the natural phenomena found in these exotic lands that Spain now claimed. They passed their excitement on to a generation of European readers.

Oviedo made a strong claim for the rightness of Spain's presence in the New World by suggesting that the natives were like stepmothers to the land instead of the true mother that the Spanish were. These portrayals had long infuriated Las Casas, who said of Oviedo's work that it contained more lies than pages—hardly a just assessment, for Oviedo filled his magnum opus with keen observations about the Caribbean world he carefully studied.

Throughout his decades of service in the colonial administration, he assiduously took notes on the novelties around him, sketching plants, animals, people, and their artifacts. He recognized very quickly that the Spanish had found a true new world.

When Oviedo gets down to cases, he gives exact descriptions of wasps, bees, ants, snakes, and alligators along with details of strange animals and extravagant edibles. Much of his amazement at the fecundity of the Indies stemmed from the dramatic difference between the tropics and the familiar European temperate zone. To other Spanish, the wetness made the country unfit for habitation. The dampness had impressed itself on Columbus, who explained the shallow roots of great trees as stemming from their steady supply of water from the frequent rain. According to Oviedo, when Queen Isabella heard this she exclaimed that "this land, where the trees are not firmly rooted, must produce men of little truthfulness and less constancy." He elaborated on Isabella's conjecture with facts about the laziness of the Amerindians. He even thought that some Spaniards had acquired these native habits.[16] In another century these speculations would lead to much theorizing about the influence of climate upon social customs—even political institutions.

Oviedo had a strong sense of writing a history while the history-makers still lived. He wasn't poring over ancient texts to reconstruct the Battle of Thermopylae but rather talking to the actual participants of events. Later generations, he realized, would find them as fascinating as his contemporaries did the Greeks' brave stand against the Persians. He named Pliny, the first-century Roman historian, his model, as well he might since Pliny had composed a compendium of ancient knowledge that included botany, zoology, metallurgy, and pharmacology. Yet he wrote to dazzle his readers with the wonders of a world that the ancients knew nothing about. He could hardly contain himself when describing the abundance of lofty trees, high and fertile mountains, extrav-

agantly colored birds, seafood delicacies, beautiful rivers, delicious tropical fruit, fragrant flowers—things he said would have
astounded Ulysses and Theseus.

Tropical flowers delighted with their brilliant colors, intricate
designs, and the sensual texture of their petals. In this Oviedo
followed the model of Columbus when he rhapsodized over the
intensity of that beauty in the New World in the diary of his first
voyage. When words failed to convey what he had seen, Oviedo
drew pictures of the people, landscape, plants, and animals that
filled his pages. Seventy-six of his drawings have survived, tes-
taments to the staggering task of making his countrymen back
home believe what they could scarcely imagine. Obsessed with
accuracy, he supplied maps for specifying locations.[17]

In order to improve his narrative of Cortés's conquest of Mexico, Oviedo interviewed Juan Cano, Montezuma's Spanish son-
in-law. He also learned firsthand about expeditions from Diego
de Almagro, Francisco Pizarro, Andrés de Urdaneta, Hernando
de Soto, and Juan Ponce de León. Alvar Nuñez Cabeza de Vaca
told him about the trek across the upper Gulf coast that started
in Tampa Bay in 1528 with three hundred men and ended four
years later in Mexico City with four survivors. Oviedo recorded
the impressions of those who lived through Francisco de Ore-
llana's spectacular voyage down the Amazon.[18] Oviedo had the
instincts of an investigative reporter, and he was proud of getting
his facts from participants, pointing out in the prologue to his
history how many other histories had been more attuned to a fine
style than to veracity, "for just as a blind man cannot distinguish
colors, so the man who is not there is unable to attest."[19]

In 1526 Oviedo published at his own expense his *Sumario,* the
first account of the New World that satisfied the learned readers
eager for an authentic picture. The subsequent Italian, French,
and English publications of the *Sumario* were different versions
rather than translations, because the unauthorized publishers took

many liberties with their texts.²⁰ Six years later, Charles I named Oviedo royal chronicler of the Indies, as well as garrison commander of the royal fortress at Santo Domingo, posts he held until his death twenty-five years later. Setting up housekeeping in Santo Domingo with a third wife, Oviedo continued to write numerous letters of advice to the king.²¹

Fitted out with a library as well as a zoo in the fortress, Oviedo was now ready to begin his monumental *General and Natural History of the Indies.* An inveterate rewriter, he still kept polishing his *Sumario* as he wrote and rewrote the grand history of which it was a summary. In the interval between publishing the *Sumario* and the *General and Natural History,* Oviedo learned where he had erred earlier.²² With evident reluctance he gave up the story of the women warriors of the Amazon Valley who burned off their right breasts to improve their accuracy with bow and arrows.

Many of Oviedo's contemporaries worked hard to place the natural phenomena of the New World into Pliny's schema. Finding similarities at first came easy to casual European observers. The puma looked like the lion; maybe the llama was a kind of camel, the jaguar a tiger, the alpaca a sheep. But the exactness of the reports from Oviedo and others—along with the creatures themselves—forced their readers to accept that the New World was indisputably new. Its flora and fauna could not be made to fit into the botanical categories that Europeans used. "They are so different," wrote one expert, "that to seek to reduce them to species known in Europe will mean having to call an egg a chestnut."²³ If this was so for plants and animals, the problem was compounded when it came to the human occupants whose unstructured way of living offended and confused the Spaniards.

Considering the dryness of the Iberian Peninsula, it's not surprising that the Spaniards were overwhelmed by the vibrant appearance and sound of the tropics—the thousand shades of green and the constant melody from chirping insects, amphibians, and birds.[24] The tropics also possessed what Europeans longed for: sweet things like sugar. And there were chilies, avocados, beans, potatoes, cocoa, tomatoes, papayas, guavas, and peanuts. Oviedo and Las Casas disagreed on the peanut. The aristocratic Oviedo considered it fit only for natives and children, while Las Casas found it more tasty than any Spanish nut.[25]

The pineapple, on the other hand, thrilled Oviedo. He struggled to convey its texture, its taste, and its appearance, which had struck the Spaniards who first saw it as being like a pinecone. Without a general classification system for plants or animals, Oviedo puzzled about whether or not resemblance should be the measure, but there was no uncertainty about its beauty. He enumerated all the European countries that lacked anything comparable while extolling its fragrance and taste with instructions on how to peel and slice it.[26]

Spaniards saw the pineapple first in a striking setting during Columbus's second voyage, when they happened upon preparations for a feast on the island of Guadeloupe. They found a pot of body parts boiling, with stacks of vegetables and fruits, among them pineapples, nearby! Weeks at sea spoiled pineapples sent home, so Europeans were introduced to them as dried fruit. Soon the pineapple became a sign of royal luxury and a symbol of luxurious hospitality and fecundity. Oviedo declared God the first gardener when he described the beautiful trees that grew wild, but this comment shortchanged the natives' skill at agronomy. The pineapple actually came from Brazil and Uruguay. Seafaring Caribs and Tainos had carried transplants in their canoes, to be cultivated on their islands. And no wonder, considering its delicious taste.[27]

The Indian way of life both attracted and disgusted the Spaniards. The exuberant vegetation and plentiful fish and fowl in the tropical zone explained why the natives appeared not to work. Their physicians conducted their idolatries and sacrifices, but

Oviedo remarked favorably that they were pretty good diagnosticians when they prescribed their herbal medicines. Not one to
overlook details Oviedo described how the Cueva Indians in the
Panama area "go about naked, and on their male member wear a
shell or a wooden tube, with their testicles out." The women, he
reported, wore petticoats of cotton blanketing, down to the ankles
for noblewomen, who also wore a gold bar under their breasts ("of
which they are very proud") to keep them high. Explaining the
servile status of most Cueva, Oviedo could not resist expressing a
bit of admiration for the smooth operation of the relay system of
litter-bearers that carried their chief about.[28]

Cannibalism and nakedness were enough to damn the Amerindians in the eyes of most Europeans. Even though it was clear
from Oviedo's reports that the natives had complex mores and
rituals, their strangeness inhibited respect. When the first part of
Oviedo's history appeared in 1537, it outraged Las Casas because
of Oviedo's hypocrisy in pretending to be an objective observer of
the Indians when he held them as slaves. He succeeded in blocking publication of the second and third parts of Oviedo's massive work, which contained a much more sympathetic view of the
Indians, the result of Oviedo's changing opinion. Because they
were unpublished until 1535, these later views remained buried in
Oviedo's manuscript, leaving his published texts to support his
reputation as an Indian-hater.[29]

During the two years that Las Casas spent in Spain, he was
able to persuade Cardinal Francisco Jiménez de Cisneros, the most
prominent prelate in Spain, to name him Protector of the Indians. His report to the king on remedies for abuses of the Amerindians passed through the hands of Erasmus to Thomas More
who used some of the material in his famous *Utopia*.[30] Most significantly, he secured the king's approval to found an exemplary
community in the northern coast of Venezuela. Las Casas has the
distinction, perhaps dubious, of being the first European to see

the empty stretches of land in the New World as a canvas upon which to paint a brighter future for humanity. Dozens would hearken to this possibility in the next three and a half centuries with rare success.[31]

Despite a promising beginning in recruiting settlers, Las Casas failed to secure their safety. Slave-traders poisoned the natives against all Spanish, and the enterprise failed within a year. Chastened by this setback, Las Casas became a mendicant friar, joining the Dominican order that he had long admired. He withdrew from the battle with the *encomenderos*, shifting his attention to Nicaragua and recently conquered Guatemala, where he succeeded in replacing aggressive pacification of the Indians with patient conversion. This won him the support of Pope Paul III, who issued a bull threatening with excommunication those who enslaved Indians.[32]

In the 1530s Las Casas began writing his great opus, *The History of the Indies*. He stipulated in his will that it not be published until forty years after his death, but he extracted from it *Brief Relations of the Destruction of the Indies*, a polemic fiery enough to fuel a long campaign against what he considered the calumnies of Indian-haters. *Brief Relations* circulated widely in Spain and through English, French, Flemish, German, and Latin translations. Other Europeans drew on Las Casas's disparagement of the conquistadors for their own purposes, laying the basis for the so-called Black Legend of Spanish atrocities in the New World. Particularly popular among Protestants, the Black Legend was to haunt Spain for generations.[33]

When Las Casas was trying to explain to King Ferdinand the horrifying slaughter of the Indians, he used the word "devastated" to make his point. Normally, he wrote, the word refers to a great loss, but when used about the indigenous people of the New World, "we mean . . . that we have seen them as filled with people as a hive is filled with bees, and that now they are all devastated, the Spaniards having...slain them all...so that nothing is left of the villages but stark walls...and all the people have died." Later he specified how the four million natives of Hispaniola had dropped to two hundred, with similar ratios in Cuba, Puerto Rico, Jamaica, and the mainland areas.[34]

Where Oviedo celebrated the heroic triumphs of the Spaniards in the New World, Las Casas detailed the terror the Spaniards spread, working with weapons, horses, and dogs. They kill as for a joke, he maintained, making Indian men and women strangers in their own land. The conquistadors were like "famished wolves, tigers, or lions" falling on a flock of gentle sheep.[35] And this behavior of the Spanish under the cloak of spreading Christianity doomed any true conversions among the native people, Las Casas stressed.

Concerns about the justice of the conquest of New World lands and the treatment of its people worried the Spanish monarchs, so much so that Charles I called an end to future conquests until there could be a resolution to these issues. Then he summoned a group of fourteen theologians and counselors to assess the opinions that Las Casas and the Aristotelian scholar Juan Ginés Sepúlveda presented to them. The two learned men debated two propositions: whether it was "lawful for the King of Spain to wage war on the Indians, before preaching the faith to them" and was it right "to subject them to his rule, so that afterwards they may be more easily instructed in the faith?"[36] Both Las Casas and Sepúlveda were deeply versed in theology and civil law, but Las Casas had the advantage of actually knowing the Taino, Arawak, and Mayan men and women firsthand. Sepúlveda had to rely upon Oviedo's portrayal of the Indians as bestial people who practiced sodomy, idolatry, human sacrifice, and cannibalism.

When Sepúlveda invoked Aristotle's concept of natural slaves, Las Casas dismissed the Greek philosopher as not being a Christian. Rather he was "a pagan now burning in hell" and hence not bound to his fellow man by the injunctions of Jesus Christ. Recognizing a brilliant polemicist, Sepúlveda described Las Casas as "most subtle, most vigilant, and most fluent, compared with whom the Ulysses of Homer was inert and stuttering."[37]

Las Casas shrewdly saw that exaggerating the differences between themselves and Amerindians made it easier for the Spanish to maintain that the native men and women were somehow less than human. Restoring their humanity became his principal task, taking on the charges of bestiality, indigence, and irrationality that Oviedo and others had put into circulation. Because many of the arguments among the disputants relied on facts rather than doctrine, Oviedo and Las Casas gave exquisite attention to descriptions of native behavior. Their nakedness, lack of shame, and possible communion with the devil were carefully scrutinized. "To give substance to the greatness of the Indies," Las Casas said, "one would need all the eloquence of Demosthenes and the hand of Cicero."[38]

To bring the Indians into the more benign category of being like other historic outsiders, Las Casas carefully explored their cultural practices, giving attention to their mechanical arts and daily routines. He instanced their capacity to adopt superior Western ways as proof of their readiness for conversion. The role of language in mediating the relations between the Spaniards and their new subjects and slaves intrigued Oviedo as well. He agreed that Indian languages were not crude and only appeared that way because they were foreign.

Las Casas noted that the Greeks called others barbarians because they didn't speak Greek. The word "barbarian" came from "babbler." Oviedo pointed to the difficulty a Basque speaker would have "speaking German or Arabic, or any other unfamiliar language." Alas, these shared observations did not keep other Europeans from insisting for another century that the Indians spoke gibberish.[39]

In the immediate moment, *Brief Relations* helped Las Casas and his fellow defenders of the Indians to secure passage of the New Laws that somewhat mitigated the harshness of the *encomienda* system. Despite the crown's desire to deal justly with its native subjects—law reforms had been attempted three times before— no one was willing to undermine the entire Spanish enterprise in the New World, which many believed would have been the consequence if Las Casas and his fellow friars had triumphed.[40]

The rapacity of all the Europeans in the Caribbean was so marked that contemporaries coined the phrase "beyond the line" to indicate the absence of all restraint there. Our knowledge of this fact robs us of Las Casas's perspective. He, after all, arrived in the New World in the first decade after Columbus's voyage of discovery, when he could reasonably expect that relations with the Indians would improve once the Catholic hierarchy and the Spanish crown understood what was happening.

In his last years Las Casas changed approaches: he converted Indian men and women, railed against their masters, lobbied at court, used his authority as a bishop to force better behavior on the settlers, and wrote fierce denunciations. He kept trying different tactics in the vain hope that he would find the right one. His initial impulse was reasonable; his persistence perhaps blind, but Las Casas was not naïve. There was every reason to think that witnesses to the violence could move church and crown officials to order reform of the dreadful system.

Las Casas was not alone in studying Amerindian ways of life. While soldiers, royal officials, and fortune hunters who composed the charter membership of Spain's new empire were mainly impressed by the nakedness and indolence of the native people, the missionaries, as distinct from the priests who served the Spanish, saw through this surface distinction. The Franciscan Bernardo de Sahagún also presumed that the Amerindians had a culture worth studying and devoted his life to doing so. Having trained at Salamanca, then the center of humanist thought in Spain, Sahagún brought his great learning to Mexico shortly after Cortés's conquest and remained to serve a half century in a convent near Mexico City.

Driven by a profound evangelical passion, Sahagún recognized after a decade spent in the New World that the mass baptisms that had marked the first conversion campaign had not deeply touched the converts. He had a strong sense of Christianity as the religion of the Word, and that converts had to understand Jesus's good news, his gospel, before any real conversion could take place.[41]

A natural linguist, Sahagún learned Nahuatl, the language of the Aztecs, and used it to penetrate Aztec mores, habits, and cosmology. He interviewed hundreds of Aztec elders, including women, to form a picture of the world as the Aztecs saw it. "Picture" is the right word, for Sahagún taught several generations of Aztecs to depict their world in drawings.

Sahagún's native draftsmen in Mexico drew unusual life forms such as the axolotl, a salamander-like amphibian whose capacity to regenerate parts of its body has made it a staple of biological studies ever since. A unique specimen from the New World, the axolotl suggests that once human beings had the ability to regenerate limbs.[42]

Sahagún's fluency in Nahuatl enabled him to reach Aztec converts and train them to interview their elders. It was an amazing and unique investigative enterprise capped by a revelation of the Aztecs' interpretation of the conquest of Mexico. Like Las Casas, Sahagún was tilting against the wishes of Spanish settlers, who opposed efforts to treat the conquered people with humanity, which they astutely saw as an impediment to their exploitation. But if to know is to understand, Sahagún's work succeeded in investing the conquered people with dignity. Motivated by devotion, his life reveals him as a bold thinker and imaginative writer.

While Sahagún translated the Gospels into Nahuatl, less predictably he preserved the formal orations of Aztec leaders in Spanish and Latin. In this he ran the risk of appearing to justify their idolatry.

Sahagún did not work alone. The Franciscans founded a college in 1536 through which they recruited and trained native leaders to carry on an extensive evangelical effort. Nothing escaped Sahagún's investigative spirit, from Aztec family customs and recipes to herbs, native wildlife, and the network of roads and canals in Mexico City. In addition to his ecclesiastical duties, he taught Aztecs to read and write Spanish and Latin, as well as to draw in the European fashion.

The fruit of Sahagún's intense and extensive investigations appeared in his twelve-volume *Historia General de las Cosas de la Nueva España* (*General History of the Things of New Spain*), which has survived in manuscript as the 2,400-page Florentine Codex. His students, recalling the lives of their forebears, produced 2,468 illustrations depicting the Aztecs fishing, farming, making music, and preparing for war, along with Aztec women preparing medicines. The small, flat drawings of warriors, pharmacologists, musicians, priests, and laborers have become a feature of American textbooks since their English publication in 1982. A labor of forty-five years, the Codex set a standard for completeness and accuracy that has rarely been equaled since, earning Sahagún the title of "the father of ethnography."

The Amerindian culture that Las Casas and Sahagún studied during their long careers did not remain the same. With each passing year more of them succumbed to European diseases. The survivors suffered from the shock of seeing those around them die. Like self-fulfilling prophets, the Spanish who held the indigenous people in contempt promoted contemptible behavior by disrupting their work, worship, and mores. Meanwhile a whole new population of those of mixed blood burgeoned. Soon the Spanish had coined precise terms to denote the exact mixtures of African, European, and Indian heritage in a person.

Much of the controversy about the mores of the native peoples swirled around the hot issues of the Aztecs' human sacrifices and the cannibalism of the Caribs in the Lesser Antilles. Las Casas, always an ingenious debater, appealed to the biblical account that God commanded Abraham to sacrifice his only son. More effectively, he justified human sacrifice as integral to the Aztecs' worldview. With a moral imagination unequaled in his time, Las Casas sought an accessible humanity in the Indian ways.

Well could Las Casas say with the Roman playwright Terence, "nothing human is alien to me," although for Las Casas this truth was grounded in Christian love. His core convictions went back to the ones he worked out in 1522 when he joined the Dominican order. The Indians, as well as African slaves, he believed, must be embraced in the gospel concept of "the poor" to whom Jesus ministered. They were deserving of sympathy and comfort, not exploitation and denunciation.[43]

Philip II succeeded to the Spanish throne after the death of his father, Charles I, in 1554. In his debate with Las Casas, Sepúlveda had been so blunt about the Spanish treatment of the native people that Philip banned the publication of his argument. But Sepúlveda won the day because he convinced Philip's generation that the Amerindians were "lazy and vicious, melancholic cowardly, and in general a lying, shiftless people."[44] By 1566, the year of Las Casas's death, Philip II was ready to start issuing licenses for expeditions of discovery. At best, Las Casas and the many other clerics and laymen who rhetorically flayed the conquistadors for their cruelties had a moderating influence on the Spanish frontier societies, awash as they were with greedy, rapacious settlers whom distance had liberated from the mores of their homeland.

The religious scruples of the Spaniards demanded that they have reasons for enslaving Indians. Debasing their intelligence and habits became a standard way of justifying their physical exploitation. More subtle were the explanations of who they might be since there was no mention of them in the Bible. The idea of some who thought the indigenous people might be members of the lost tribes of Israel was an identification with consequences, for if true they could be punished for lapsing from the true faith. Many others saw them as creatures of the devil, a status that would sanction anything done to them. Still others held out hope of converting them to Christianity. Spaniards holding this view described the *encomienda* system as a form of tutelage, a holding of the Indians "in trust" while they caught up to their more civilized masters by working for them.

One scholar, Gregorio García, had written a massive tome in 1607 reviewing the possibilities that the natives whom Columbus had encountered might be Jews, Carthaginians, Greeks, Romans, Phoenicians, Egyptians, Ethiopians, French, Cambrians, Finns, Frisians, or Scythians. He finally concluded that everything in the New World was essentially different. What I say about various items, he wrote, "I will say of a thousand varieties of birds and fowls that have never been known by either name or appearance, nor is there any memory of them in the Latins or Greeks, nor in any notion of our world over here." For some this fact suggested that there was more than one site where human life began. For those who scorned polygenesis, the search continued for biblical signs in the New World. Spanish friars cooperated in this effort by finding traces of a flood in Mexico and Peru, as well as crosses strewn throughout Spanish America.[45] It took a century of speculation to conclude that nothing in the New World could be fitted into European knowledge based on classical and biblical texts.

Greed, ambition, and fractious infighting dominated the stories of the Spanish Empire. Yet the protracted debate about Spanish obligations to the native people prodded reformers into learning about their ways the better to defend them from oppression. The careers of Sahagún, Las Casas, and Oviedo show how profound an impact the New World had upon those given to reflection. Arriving as young men, they recognized that the future would differ irrevocably from the pre-Columbian past. They wrote voluminously about the people, the plants, the animals, and the landscapes that spread out before their astonished eyes. They gave their countrymen back home—and their descendants—permanent sources of information about the mysterious antipodes—enough material to sustain generations pondering the joining of the old and new worlds. Alexander Humboldt, the great nineteenth-century explorer, lamented the difficulty of sorting out "the facts attested to by a crowd of eyewitnesses, all of whom were enemies of each other." But without their accounts, a record of the people of the Western Hemisphere when they encountered the strangers from the East would not exist, nor would the stimulation of the Europeans' imagination have been so powerful.

What gets lost in the dramatic recounting of Columbus's four voyages is the curiosity that his discoveries provoked. The lure of riches, new territorial possessions, and triumph over rivals probably furnished the motivation of the sponsors of the subsequent succession of expeditions. Yet the intellectual consequences of Spain's venture into the Caribbean have long outlasted its empire. The novelties of the Western Hemisphere nudged Europeans toward modern ways of thinking about their planet. Slowly the age-old concern with acquiring wisdom through contemplation was pushed aside in the search for mundane details about the earth and its peoples. Old ways of knowing were turned upside down.

A passion for collecting information through observations, measurements, descriptions, and depictions of new phenomena grew stronger, though it took successive generations to assess the new data and generate new hypotheses about their meaning. This awakened curiosity opened up the doors of inquiry to less educated amateurs, once excluded from the closed circles of the Scholastics and humanists. Europeans turned their attention to inspecting, comparing, categorizing, and explaining natural phenomena. This enterprise began with authors, not adventurers.

CHAPTER TWO

DISCOVERY OF THE OTHER HALF OF THE GLOBE

Painting of the Strait of Magellan during Magellan's 1520 circumnavigation of the world by British artist Oswald Walter Brierly (1817–1894).

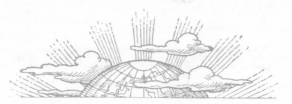

Columbus, adventurous as he was, never gave up his belief that he had reached the outskirts of Asia, despite three additional exploratory trips in the Caribbean. Nor were his fellow Europeans any more prepared for the challenges to what they knew or thought that they knew. Old paradigms only fade when new ones with different factual foundations have been established. This was particularly so for a society in which nothing was more novel than novelty itself.

It took other navigators returning with wide-eyed enthusiasm about the unusual American people, their habitats, the landscape, and the flora and fauna to challenge Columbus's assumptions. The Florentine Amerigo Vespucci, who participated in several exploratory Portuguese voyages along the east coast of Brazil, determined that South America was indeed a new continent. The popularity of his 1502 travel account spread this idea throughout Europe. Five years later a German cartographer, Martin Waldseemuller, was working on a new map, and, having read Vespucci's account, chose the name "America" for the continents of the Western Hemisphere.

Meanwhile Columbus continued to look for a strait that would let him sail to the Indian Ocean. On his fourth and last voyage, in 1502, he explored the coast of present day Honduras, Nicaragua, and Coast Rica, setting up a garrison in Panama. Discouraged by his failure to find a way out of the Gulf of Mexico, he set sail for Hispaniola, where his sworn enemy, Nicolás de Ovando y Cáceres, had been installed as his successor to the governorship. Stranded in Jamaica, Columbus was able to persuade the natives to continue provisioning his men by predicting a lunar eclipse.

With a bit of secret help, in 1504 he finally got back to Spain, where he died two years later. Making discoveries passed to other eager hands. The rest of the globe was yet to be explored. The vast ocean west of the Americas contained a multitude of islands with their own unique inhabitants and habitats.

Inspired by Columbus's voyages, Vasco Núñez de Balboa sailed for Spain's New World empire in 1500. By 1510 he founded the first permanent settlement on continental America in what today is Colombia. With the characteristic ambition of the conquistadors, Balboa led several expeditions to search the area for gold and natives to enslave. All the while, he maneuvered among various Amerindian tribes, defeating some, turning others into allies, and converting several chiefs into Christians. Without scruples, but with a great deal of charm and good sense, Balboa conquered the Isthmus of Panama for King Ferdinand, now, after the death of Isabella, the sole ruler of Spain. Spurred by rumors of a great sea to the west of his settlement, he set off to find it in 1513 with 190 Spaniards, some native guides, and a pack of dogs.

Fighting hostile natives through jungles, Balboa and a band of hardy survivors reached the summit of the mountains rising above the Chucunaque River. From there they beheld a great body of water in the distance. Descending to the water itself, Balboa joyfully raised his sword and an image of the Virgin Mary and claimed the sea and all the lands it washed upon for his king. A contingent of officials, including the chronicler Gonzalo Fernández de Oviedo with his wife and children, arrived from Spain to take charge of these new possessions, kicking Balboa out. Ever the conquistador, Balboa continued to explore the isthmus and plunder its native inhabitants. Finally running athwart the new governor, he was condemned as insubordinate and killed in 1519 near the settlement he had started.[1]

Eight months after Balboa's execution, Ferdinand Magellan set off from Spain in a five-ship expedition optimistically named the Armada of Molucca. Several mariners had convinced Charles I that the Spice Islands lay on the Spanish side of the pope's dividing line. Magellan would do what had eluded Columbus: find a western route to the Indies. It would also fall to Magellan to name the great ocean that Balboa had discovered. Sailing into it on a calm day, he named this most turbulent of seas the Pacific Ocean. It had taken Columbus seven years to get support for his first voyage. Magellan had to wait less than a month for the go-ahead, with Spanish officials eager to get to the sources of nutmeg, cinnamon, cloves, and mace before the Portuguese. He did not yet know that the Portuguese had already reached the Indies, so shrouded in mystery was Vasco de Gama's success. Eventually the Spanish too became secretive, restricting both the flow of information and access to their empire.

Bartolomé de Las Casas, fortuitous again, had been at court the day that Magellan, carrying a colored globe, presented his case to Charles I. He later engaged in conversation with Magellan, who told him of a strait that would take his fleet directly from the Atlantic to the southern sea that Balboa had discovered. Magellan had probably learned about the strait from the navigational charts of Martin Behaim, a German astronomer and cartographer at the Portuguese court.

So precious did Magellan consider this information that he memorized the charts rather than rely on a drawing that someone could steal. The mysterious passage was crucial to the enterprise for, when Magellan set sail in 1520, Europeans had yet to locate the tip of South America or even be certain that it had one. The strait, which later carried Magellan's name, offered the only known access to the Pacific from the Atlantic, even though it turned out to be fraught with perils.[2]

In the age of sail, a navigator needed to know more than the lay of the sea; he had to understand wind and current patterns. Las Casas marveled at the confidence Magellan displayed, especially considering that the ancient historian Pliny had asserted that the seas were filled with monsters and fantastical beings like troglodytes. The first voyagers had no way of knowing whether they'd encounter three-hundred-foot-long eels or meet people with horse's feet and dog's heads, as had been reported.[3] At a practical level, it was even harder to stock enough supplies to sustain a crew with edible food and fresh water.

Magellan put together his crew of Greeks, North Africans, Sicilians, French, Basques, Flemish, and Britons unmindful that one of his good friends, Francisco Serrão, a Portuguese squadron leader, was already settled in Banda in the Indonesian archipelago.[4] The most important crew member he chose was Antonio Pigafetta, an Italian aristocrat distinguished by his intellectual curiosity.

Pigafetta, slated to become one of the great chroniclers in the age of exploration, joined Magellan's epic voyage for the express purpose of making a full record of the first circumnavigation of the globe. He had rushed to Spain to find a berth in the fleet of this extraordinary venture because he was thrilled by the idea of sailing west to find the Indies, even accepting a minor position in order to see firsthand the mysterious places that lay between the Iberian Peninsula and the Moluccas.

Magellan saw in the twenty-seven-year-old a writer talented enough to deliver a full account of the venture. He wanted more than a pilot's log, which would also be kept. He desired a narrative that would rival everybody's favorite: *The Travels of Marco Polo*. Pigafetta, a fine stylist and a formidably intelligent man, well might write a contender to it.[5]

In September of 1519, with 270 men in his five-ship fleet, Magellan set sail. One way to follow this journey is to track the dwindling of the fleet. First the *Santiago* disappeared, presumably sunk, while the pilots were searching for the strait that would take the fleet to the Pacific in the fall of 1520. Soon after, mutineers took the *San Antonio* back to Spain. The *Trinidad, Concepción,* and *Victoria* continued, passing through the strait to enter the Pacific Ocean on the last day of November.

After battles in the Philippines, there weren't enough sailors left to man three ships, so the fleet leaders decided to abandon the *Concepción* in May 1521. The *Trinidad* and *Victoria* reached Borneo in July and arrived in the Moluccas in November. The next month the *Trinidad* broke up in a storm after the Portuguese had seized its crew, leaving the *Victoria* to complete the circling of the globe, arriving back in Spain in September 1522 with eighteen survivors.[6]

Pigafetta reported it all. He wrote about stunning novelties like men with tattoos all over their bodies, pearls the size of hen's eggs, and penguins—birds with black-and-white feathers that had to be skinned rather than plucked. He had genuine concern for the people he met and the languages they spoke, and he collected word lists from the natives of Brazil, Patagonia, the Philippines, and groups in the East Indies. While he included tall tales, he mainly chronicled what he had seen himself, detailing the routes the fleet took, its stops, the topography along the way, and the appearance and conduct of the indigenous people they encountered.

Less concerned with the politics of the voyage, which were fierce, Pigafetta gave short shrift to the details of the mutinies Magellan faced. He did acknowledge that the captain-general was well hated, a fact he ascribed to Magellan's being Portuguese and his foes a bunch of envious Spanish.[7] Magellan had become a Spanish citizen so that he could serve the Spanish crown in its exploratory voyages. Being Italian, Pigafetta could take a lofty attitude toward the enmity between the Iberian rivals.

There were understandable reasons, though, for some of the crew's hatred for Magellan. Among the many formal orders that Charles I gave Magellan was to keep the crew away from native women. Very quickly the alluring women along the South American coast undermined compliance with this instruction. The fleet stayed at Port San Julián, just north of the as-yet-undiscovered strait, for five months in the spring and summer of 1520. Crew members spent their nights on the beach with their native companions. Reluctant to leave the women behind, some sailors even hid them on the ships. A suspicious Magellan quickly found and dispatched the stowaways. When the master of one of the ships was caught sexually involved with a cabin boy, Magellan had him court-martialed, although homosexual relations were common on long voyages. Quickly convicted, the man was executed, a punishment that added fuel to the crew's fear and dislike of their commander.[8]

The sexual habits of the people encountered as the fleet moved down the coast of South America and through the island communities of the Pacific could not but have intrigued Europeans used to strict moral codes. They certainly intrigued Pigafetta. Everywhere women were dominated by men, but both men and women were much freer about expressing their desires, he reported. Unattached women were often offered to the men of the fleet, but rarely wives. With these personal details, Pigafetta's chronicle augmented the accounts of Christopher Columbus and Amerigo Vespucci, opening for Europeans new windows on the diverse ways societies handle the explosive power of erotic passion, always and everywhere a potential threat to order.

The nakedness of Amerindians and the occupants of the islands of the Pacific both titillated and disgusted Europeans. Sailors were a notoriously lewd bunch, so they interpreted the nudity of the women they met as an invitation. Pigafetta, having become aware of the moral systems of the groups he was studying, realized that they just didn't prize virginity.

Vespucci had also written on inhabitants' sexual mores in *Mundus Novus* and *Lettera al Soderini*, published in 1502 and 1504, in which he described, accompanied by provocative drawings, naked men and women enjoying sexual freedom as well as the taste of human flesh. He didn't moralize about the Indians, dispassionately discussing their unfamiliar mores and handsome bodies. Vespucci spent twenty-seven days living with Amerindians whose way of life he contemplated, apparently never tempted to enslave them or render prejudicial verdicts of their differences.

There is much controversy about whether Vespucci, a Florentine merchant who also became a skilled pilot, made one or four voyages to the New World between 1497 and 1504, or whether he actually wrote the letters that made his name famous. Whoever the author, the books found many readers. By 1508 the Vespucci letters had been published in thirteen cities in Latin, German, Dutch, Italian, Czech, and French.[9]

In the Vespucian tradition, Pigafetta carefully noted the various forms of near nakedness he observed as the Magellan expedition proceeded. He elaborated on the jewelry with which nude women and men often adorned themselves. He was also attentive to body painting, body piercing, and tattooing. Almost all the men, he reported of one South American group, "have three holes pierced in the lower lip, holding round stones, one finger or thereabouts in length and hanging down outside."[10] They also plucked out their body hair.

Most fascinating to Pigafetta were the men on Cebu, in the Philippines, who pierced their penises with a gold bolt with a hole in the middle for urinating. Incredulous, he kept checking out penises and querying the men about how intercourse was possible.

Magellan became a successful proselytizer in the Philippines, converting several important leaders. One of them got Magellan to attack his enemy. Taking three boatloads of sixty men, Magellan landed in the enemy's territory. Alas, he was outmanned and outmaneuvered, his force soon overrun. Most of his party took flight, leaving him as the sole target of hundreds of enraged fighters.

Pigafetta was in the invading force and vividly recorded the disaster that took place. "Recognizing the captain, so many turned upon him that they knocked his helmet off his head twice, but he always stood firm like a good knight . . . an Indian hurled a bamboo spear into the captain's face. . . . When the natives saw that, they all hurled themselves upon him. . . . Immediately they rushed upon him with iron and bamboo spears and with their cutlasses, until they killed our mirror, our light, our comfort, and our true guide." Pigafetta penned a memorial to Magellan, praising his constancy and endurance, adding that no other had "so much natural talent nor the boldness nor the knowledge to sail around the world, as he had almost already accomplished."[11]

The newly elected commander of the remaining ships in the fleet, Juan Sebastián del Cano, decided to sail on, desperate to reach the safety of the high seas after another encounter with hostile Filipinos. Pigafetta reported on this horrendous event and then treated his readers to a list of Filipino words.[12] His lists of words in Malay and other languages became a great source for philologists later, but his concern for the lists points more to his reluctance to dwell on untoward happenings in a voyage that was already packed with disasters.

By the time he reached the Indies, Pigafetta was prepared for the phallic decoration of tiny bells that he saw. The men there told him that their women loved the sound of the bells tinkling inside them during intercourse. Needless to say, such details opened the European imagination to a much more complex understanding of social mores that would in time flower into the full-blown discipline of ethnography.

In Java, Pigafetta learned about the practice of suttee, described by his native interpreter in glowing terms as a ceremony in which a flower-bedecked widow happily accepted her duty to join her husband's corpse on the funeral pyre. He presented his readers with another difference between them and the people of the Marianas. He was most impressed, as a young Italian nobleman, that "each one of these people lives according to his will, for they have no lords." Vespucci had made a similar observation after liv-

ing with the natives of Brazil for a month when he wrote in his
Lettera al Soderini: "Having no laws and no religious doctrine, they
live according to nature," adding, "they have among them no private property, for everything is in common.... Each is his own
master...."[13] Such an egalitarian idea appeared strange to these
Europeans imbued with hierarchical assumptions about property
and status.

The members of the expedition—now numbering 108 men—
vowed to avoid further rash distractions. They headed straight
for the Moluccas. The *Trinidad* and *Victoria* reached Borneo two
months later, after receptions both friendly and hostile in the
islands dotting the seas east of the Philippines. These visits produced new perplexities about nakedness. The people of the Philippines maintained a lively trade with the Spice Islands, which
permitted their royalty, though naked, to sit on silk cushions and
eat off porcelain and gold dishes.[14] One island near Borneo was
occupied by Muslims who, though as naked as the other natives,
adhered strictly to Muslim rules about diet and hygiene. Here
were "people of the book" behaving as heathens, at least on one
count. Later, in the eighteenth century, when Europeans explored
the Pacific more thoroughly, they had a change of heart and would
romanticize the naked Tahitians as noble savages for whom the
pleasures of the flesh were indicative of simplicity and exuberance.

The voyage's pilot, Ginés de Mafra, kept a record of the expedition, doing a much better job than Pigafetta of distinguishing
one island from another. In a twenty-first-century coda to the
Magellan circumnavigation, anthropologists today are conducting excavations made possible by "Mafra's notes."

By November, Magellan's Armada had reached the Moluccas,
two years, two months, and twenty-eight days after their departure. Remaining six weeks at their final destination of Tidore, the
officers negotiated a treaty with the potentates of the Spice Islands
and began buying spices with their remaining trade goods. The

fact that there were any goods left is something of a miracle. The holds of these ships seemed bottomless. After two years at sea, the explorers were still dispensing presents of velvet chairs, damask robes, and cubits of cloth. At one stop, the visiting dignitaries were given "some green damask silk, two cubits of red cloth, mirrors, scissors, knives, combs, and two gilt drinking cups."[15]

The ever-nosy Pigafetta found roaming around Tidore a delight. At several of the stops, he sought out the plants that produced the wonderful spices and lovingly described them. The clove tree he depicted as tall and "thick as a man's body" with leaves like those of the laurel, the cloves themselves growing at the end of twigs, ten or twenty to a cluster. "When the cloves sprout they are white, when ripe, red, and when dried, black." He went on to explain that those trees "grow only in the mountains. . . . No cloves are grown in the world except in the five mountains of those five islands." The nutmeg tree he compared to walnut trees, its fruit the same color and size as a quince. The brilliant red spice, he noted, "is wrapped about the rind of the nut, and within that is the nutmeg." Ginger, he found, on the other hand, was not a tree but a small plant that put forth reeds. Thus he was able to domesticate with comparisons these most exotic of all commodities for his European readers.[16]

The Spanish constructed on Tidore a fortified locker for the overflowing goods that they were unable to stuff into the holds of the two remaining ships. Leaving nothing to chance, four crewmen were left to protect this treasure, but to no avail. It was soon lost to the Portuguese, who took the spices as well as the men guarding them. By that time, Cano and his men learned that they had sailed 10 degrees west into the Portuguese half of the globe. They also realized that the Portuguese had already reached the Spice Islands and were prepared to defend their supremacy there.

Sailing home, the *Trinidad* was lost in a storm. The *Victoria*, the only ship to fully circle the globe, made the trip back in nine

months. Some twenty-five men died on this long voyage, and thirteen more were lost when the Portuguese seized their longboat as
it went ashore for food in the Cape Verde Islands off the west
coast of Africa. A witness described the eighteen survivors of the
earth's first circumnavigation as wan and skinny; their boat having more holes than a sieve. When the *Victoria* arrived in Spain, its
hold contained twenty-six tons of cloves, nutmegs, and cinnamon.
Sold in Antwerp a year later, the returns covered the entire costs
of the expedition plus a 6 percent profit.

Even more important than the rich lode of spices were the
firsthand observations of the people and islands of the Pacific.
Europeans were able to calculate the circumference of the globe—
24,902 miles—although it took cartographers several more decades to accept this fact. In 1550 a contemporary hailed the
circumnavigation as "the greatest event since the creation of the
world, excepting the incarnation and death of Him who created
it."[17] Much remained unknown about the globe, but astronomers,
physicists, mathematicians, and cartographers had acquired an
enthralling dimension to ponder.

Representatives of the two Iberian crowns met in 1523 to
sort out their dominions. Spain continued to trade fitfully in the
Indies, but without fueling stations like those that the Portuguese had established, it proved difficult. When Charles married
the sister of the king of Portugal, he sold his new brother-in-law
Spain's rights to the trade, even as he laid plans for adding the
Philippines to his expanding Spanish empire.[18] After the discovery in 1565 of a route with favorable winds to carry ships back east
back across the Pacific, Spain sent annual silver fleets from Acapulco on the west coast of Mexico to buy spices, porcelain, ivory,
and silk cloth in Manila. Brought back to Veracruz on the Gulf
of Mexico, most of the goods were then shipped to Spain. But by
then other players—and pirates—had entered these waters. Francis Drake sacked the Spanish silver fleet during his 1577–80 cir-

cumnavigation of the globe, enabling one of his investors, Queen
Elizabeth, to pay off her country's debt.

Pigafetta's hopes that his journal might have a scientific value
in addition to being a record of a remarkable trip were put at risk
by a puzzling anomaly. Having kept a very careful account each
day, he was confused to learn that they had reached the Cape
Verde Islands on a Thursday when, according to his reckoning,
it was Wednesday. Upon their arrival in Seville, authorities dispatched a special delegation to the pope to address this conun-
drum. It took another 350 years before a conference established
the International Date Line to mark when one day changes into
the next day.

Francisco Albo, a Greek officer who became pilot of the *Victoria*, had kept a logbook.[19] Without more than dead reckoning for
determining longitude, his notations were estimates. Another of
the fleet's officers who knew mathematics observed the celestial
bodies, employing instruments he set up whenever they were on
land. Using astronomical tables, he was able to measure the longitude of their position with what turned out to be amazing accuracy. Albo's logbook confirmed that the Spanish had indeed sailed
into Portuguese territory, an unwelcome piece of information at
the royal court when the commander, Cano, made his report.

Pigafetta believed that his account was its own treasure. He
told Charles I that he had brought back neither gold nor silver, "but
something more precious in the eyes of so great a sovereign . . . a
book written by my hand of all the things that had occurred day
to day in our voyage."[20] But he was not in a hurry to publish his
precious record of the circling of the globe, preferring instead to
make the rounds of European courts to give his story in person.

Drumming up his adventures turned Pigafetta into quite
a celebrity. After his appearance before Charles, he visited the
courts of King João of Portugal; Louise of Savoy, the mother of
Francis I of France; and Isabella d'Este of Mantua. The Vene-

tian doge arranged for him to speak before the entire Venetian College. According to one listener, the audience was "stupefied by those things that are in India."[21]

Meanwhile Maximilian Transylvanus, a young Flemish diplomat at Charles's court, interviewed some of the other eighteen survivors of the Magellan expedition and rushed an account into print in the autumn of 1522, giving Europeans the first word of the globe's first circumnavigation.[22] Polishing his manuscript, Pigafetta searched for a patron for its publication. Passing up several promising ones, he finally decided on Philipe Villiers de l'Isle-Adam, the Grand Master of the prestigious Knights of Rhodes. In early 1525 he presented Villiers with his finished account, a report that included twenty-three colored maps that Pigafetta had drawn. His *First Voyage Around the World* became a showcase for his wide-ranging inquiries across the island world of the Pacific. The book also secured a place in history for Magellan, whose enemies would have deprived him of credit as the first circumnavigator.

Magellan had been right; an account in the style of Marco Polo would ensure his voyage a place in history and in the imagination of Europeans for generations to come. For a while it looked as though Magellan's defamers would rob him of all credit for planning and executing the remarkable expedition traversing two oceans, but Pigafetta's overflowing admiration of the captain-general, conveyed in vivid prose, put the prize firmly in Magellan's possession.[23] Even more enduring, Pigafetta's book extended the range of questions about North and South America to the Pacific world whose people, scattered over hundreds of miles, introduced new models of society for Europeans to ponder.

Contemporary critics have credited Pigafetta'a chronicle with a modern sensibility for its humor and depiction of the "realistic fears, joys, and ambivalence felt by the crew."[24] For Europeans who stayed at home, Magellan's Armada of Molucca proved definitively that North and South America were not part of Asia, a conclusion bristling with consequences for those trying to preserve Christian cosmology.

The publications about the discoveries highlighted the unique quality of knowledge acquired through personal experience. Learned men in fifteenth-century Europe had studied Greek, Latin, and Hebrew so they could read ancient texts and ponder the wisdom contained therein. Columbus's voyages engendered a new appreciation of knowledge that came from examining objects and listening to witnesses. A change like this is more significant than it might seem, for observing and reporting involves intellectual traits that differ from scholarly ones.

Even Oviedo's ineptness with Latin, which caused others to smirk, proved a plus in dispersing knowledge more widely. He mocked the learned humanists in Spain by telling them that he knew things that they could never find in their books. "What I have said cannot be learnt in Salamanca, Bologna or Paris," he preened.[25] Firsthand witness privileged facts over thoughts.

The discovery of the New World aroused more than curiosity about exotic places. The existence of previously unknown places peopled with unlikely inhabitants posed intellectual puzzles for the leaders of Christendom. Of course Europeans knew that Asia and Africa were filled with people who weren't Christians, but they had known of their existence long enough to file them under the rubrics of heathens or infidels, presumably awaiting conversion.

One Italian scholar heard about Pigafetta's stories in the midst of giving lectures on Aristotle. Writing a friend, he reported that Aristotle and Averroes erred in thinking that the equator was too hot for human beings. There was indeed life on the other side of the globe. He added, for emphasis, "perhaps some of you will think that merchant is telling me a sack of lies and that he is a big liar. No, dear sirs, it is not possible, he wasn't the only one on that voyage."[26] The discoveries had an impact on humanists as well as the protectors of Christian dogma.

For those persons learning something secondhand, the trustworthiness of the witness became paramount. At first any account from the New World gained currency, but the many wild tales proffered raised a skeptical attitude that was to become integral to all empirical studies. A set of criteria for testing veracity, much like that of a courtroom grilling of witnesses, sprang up among experts forced to sort out the fantastic from the unexpectedly true. Philip II actually ordered that all extant copies of Francisco López de Gómara's general history of the Indies be destroyed when he heard of its many inaccuracies.

Philip II could not destroy all the sources of inaccurate or negative portrayals of Spanish actions in the New World. Girolamo Benzoni, an Italian merchant who went to the Caribbean in 1541, returned fifteen years later with a deep hatred for the Spanish. He poured his venom into his *History of the New World*. Illustrated with especially grisly depictions of the Spanish abusing the native people, the book went through several editions and translations. Philip II had as little luck banning a Dutch account of the same topic. When he implored the Duke of Alba to counteract what he saw as calumnies from his insurgent subjects in the Netherlands, the duke shrewdly replied that the Dutch rebels would only come back with "700 other pamphlets containing 100,000 new insolent gibes."[27]

Spain's neighbors already considered the Spanish aristocracy an especially proud and violence-prone group. By 1565 when Benzoni's history was first published, England and France were ready to challenge Spain's monopoly of settlements in the Western Hemisphere. Condemning their rival on moral grounds only strengthened their sense of entitlement to a part of the New World. Angered by the pope's map dividing the globe between Spain and Portugal, the French, Dutch, and English fumed over the power of papal bulls. As long as they were Catholic, they paused. But when England and the Netherlands became Protestant in the middle of the sixteenth century, they were no longer deterred from sending their own expeditions west.

Savvy Dutch sailors contributed to a breach in Portuguese control of the spice trade. Jan Huygen van Linschoten, a Dutch merchant who became secretary to the archbishop in the Portuguese colony of Goa, betrayed the trust of his sponsor by copying Portuguese trade routes and publishing them with translations in German and English.[28] By the end of the sixteenth century, the Dutch East India Company had successfully challenged the Portuguese monopoly of the European trade with the Indies. Soon the English arrived, making decisive a shift in the center of European commerce from its centuries-old location in the Mediterranean Sea to the Atlantic Ocean.

The treasures of the tropics that the Portuguese and Spaniards brought back had whetted the appetite of other countries facing the Atlantic while raising questions throughout Europe. What kind of plants produced the spices, vegetables, and fruits that grew along the globe's 20th parallel? How were the precious metals, perfumed woods, and silken fabrics that filled ships going back home used in their homelands? Where were the plants grown that produced bright dyes that expanded the range of colors for the textiles that adorned European bodies and interiors? Novel tastes, luxurious clothes, aromatic flowers, and delicious fruits stirred the desire to understand and possess them.

As the Europeans experienced a dozen new vivid shades of red and yellow in contrast to their own blander palette of green, beige, brown, and blue, people acquired a sense of a brighter world. Materialism itself was promoted by this new cornucopia of materials, both exotic natural objects and beautifully crafted jewelry and masks. Although materialism is often considered a moral failing today, it had a much fuller meaning in the early modern centuries. It invited a turning away from otherworldly interests and stirred people to pose fresh questions about the nature of the world they inhabited. It was the world of nature more than that of artifice that nurtured a new inquisitive spirit.

Still, like Columbus who believed he had found Asia when he died fourteen years after his landfall in the Bahamas, his contemporaries resisted many of the implications of the explorers' voyages. They deflected contradictions in their traditions and folded what was novel as smoothly as possible into what they already believed to be true about the world. They were reluctant to acknowledge new species—species without representation in Noah's Ark. They clung to mythic stories of female "Amazons" who burned off their right breasts to facilitate their ferocious archery.

In several large compendiums of information, the Spanish and Portuguese voyages merited a mere few paragraphs; in the case of one seventeen-volume work, eighty pages. Historians are a conservative lot. When they composed chronologies of world history at the end of the seventeenth century, they gave scant attention to the discovery of America. Contemporaries were also slow to take in what the New World would mean to the old one. The head of a prestigious boys' school in Nuremberg, faced with revising the school's curriculum, deliberately excluded material about the New World. He acknowledged Vespucci's reports of previously unknown continents and people, but decided that, whether true or false, they had little import. The distance and dangers of reaching the Western Hemisphere made it, he concluded, were "of no interest to geographers at all."[29]

In central Europe, the Reformation and the Turkish menace caused the most concern. English readers too wanted to learn more about the Turks than about the Amerindians. Still, the cognitive dissonance from the anomalies of the Western Hemisphere prompted the reading public to ask for more stories, more information, more pictures. Interest didn't flag; it grew as the vertigo-inducing qualities of the discoveries ebbed. Europeans began using the term "modern" in a new way. Awareness of the Americas probably promoted this lexical introduction with its sense of breaking with the norms of the past.

Painters and writers had helped to spread this new knowledge early on. Poets and artists were inspired by the New World within a year of the publication of Columbus's first letter. Drawings depicting handsome naked men and women in extravagant tropical settings adorned many posters. Painters idealized nude natives in poses like those of Greek and Roman statues. Tropical leaves made their appearance in more and more designs or in the background of scenes otherwise unconnected to the discoveries. Sometimes the awareness of the New World was subtle, as in Hieronymus Bosch's *The Garden of Earthly Delights*. Completed in the first decade of the sixteenth century, Bosch placed a telltale pineapple in his triptych adorned with luxuriant tropical vegetation. One expert has called Bosch's great painting "an erotic derangement." Others take the title literally. If so, Columbus's description of the mouth of the Orinoco River as a paradise might have prompted Bosch. Perhaps the Garden of Eden had indeed left a copy on earth.[30]

The dense vines and flowering shrubs that Columbus described in the Caribbean rain forests had truly surprised Europeans, for these places shared the same latitude as the Sahara Desert. They certainly undermined the conventional belief that rain decreased with closeness to the equator. Spaniards named a particularly wet stretch of land Venezuela because the houses that the natives built on stilts to avoid flooding reminded them of Venice!

Feathers gave a major hint that images of the Western Hemisphere had penetrated European aesthetics. From the black-and-white ones of penguins to the wheel of primary colors of macaws, birds and their feathers came to signal the New World. All of its European chroniclers took care to describe the feathers that the indigenous people used in skirts and headdresses. After Cabral sent home some parrots in 1500, maps appeared naming Brazil "Terra de Papagaios," land of parrots.

Illustrations of South American natives adorned with feathers were so common that by mid-century a befeathered woman became the symbol of both American continents. No less an authority than Francis Bacon hypothesized that the people in the New World wore feathers because their ancestors had taken refuge from the flood of Noah's time in the mountains where they were joined by birds able to fly away from the waters. Even the great empiricist resorted to biblical explanations![31]

Europeans in port cities in Spain, France, and England got to see some of the New World with their own eyes when fully outfitted Indians and strange-looking animals paraded through their streets. Columbus had brought back Tainos in 1493. Other explorers returned with members of the Arawak and Carib peoples. Cortés carried off Aztecs who were such adept jugglers and ballplayers that, after the show they put on in Valladolid, King Charles sent them on to Rome to entertain the pope.[32] Probably two thousand Indians of various tribes came to Europe in the generation after 1492, most of them as slaves. Some few achieved prominence as translators. The French explorer Jean Paulmier de Gonneville actually offered his daughter in marriage to a Carijo he brought back from Brazil.

By far the most pressing practical task for Europeans after their geographical discoveries was updating their maps. Most cartographers relied upon Ptolemy, the Greek who lived in the first century. An astrologer, astronomer, mathematician, and geographer, Ptolemy had made twenty-seven maps of the globe with an index of several thousand places accompanied by their longitudes and latitudes. His hypotheses about the planets contained astronomical calculations for the positions of the sun, moon, and planets, the rising and setting of the stars, and the eclipses of the sun and moon. He believed that the earth revolved around the sun, but greatly underestimated the size of the globe. His latitudes and longitudes were often wrong.

Scholars had studied Ptolemy's writing since the late fourteenth century; printed versions spread his influence. Ptolemaic texts raised questions about the globe, whose actual shape would not be determined until two centuries later. Ptolemy's maps that Europeans worked with featured his known world: that is, Europe, Africa, and Asia. The Pacific was unknown and the Atlantic figured as a water horizon looking west. Still he produced a system of great versatility that offered a starting point for all cartographic decisions in the sixteenth century.

Even after people read about the Portuguese and Spanish voyages, they had difficulty wrapping their minds around the size of either the Atlantic or the Pacific. Martin Behaim played a critical role in getting Europeans to accept the concept of an ocean world. He visited Nuremberg in 1490–1503; its City Council commissioned him to make a globe that he named "Erdapfel," or earth apple. With enthusiasm, he wrote, "let no one doubt the simple arrangement of the world, and that every part may be reached in ships as here is seen."[33]

Another great source of geographic knowledge came from portolans, sailing directions with charts and descriptions of harbors, designed for mariners. Starting in the middle of the fifteenth century, charts in the portolan tradition displayed the progressive stages of discovery from Cape Bojador on the west coast of Africa to China and from Labrador to the Straits of Magellan. By the beginning of the sixteenth century, cartographers could accurately draw the full outline of Africa. For southern Asia, they went as far as the Bay of Bengal. Portolan maps improved with each iteration, which was not true of printed maps based on Ptolemy.[34]

Dozens of mapmakers who had learned their craft reading Ptolemy's texts took up the challenge of reconciling his maps with the new information gleaned from the explorations out of the

Iberian Peninsula that appeared on the portolans. After Vespucci leapt to the conclusion that the continents and islands found when sailing west were parts of a new world, his publications provided the most authoritative texts about the lay of the land (and sea) in the Western Hemisphere. The most popular of all discovery literature of the time, Vespucci's text even penetrated a remote town in northeast France where cartographer Waldseemuller was working on a map.

Waldseemuller received a copy of Vespucci's books at the same time that his patron got hold of some Portuguese portolans, which prompted his calling the continents in the Western Hemisphere America."[35] The popularity of Waldseemuller's 1507 map carried the new designation to a receptive public. The reputation of Columbus might have sunk altogether after that had it not been for the exertions of his son and the admiration that suffused Oviedo's great work on the Spanish Indies. For both Columbus and Magellan, writers played a critical role in sustaining their rightful place in history.

In 1602 Matteo Ricci, a proselytizing Italian Jesuit, made a map for the Chinese emperor based on Waldseemuller's. It showed the Chinese where their huge land fit into the larger scheme of the earth. Designed for display on six panels, Ricci's map stretched twelve and half feet long and rose five and a half feet tall. Ricci drew the map to flatter the Chinese, whom he was trying to convert, but he also wanted to demonstrate the superiority of Western knowledge about the world from which the Chinese emperors of the Middle Kingdom had elected to isolate themselves.[36]

Magellan's circumnavigation added the west coast of South America and the island clusters scattered across the Pacific to the places coming under observation. Officials in Seville gave one cartographer, Diego Ribeiro, access to the master records of discoveries from which he produced, in 1527, a beautiful map tracking the *Victoria*, the lone ship from Magellan's fleet that successfully completed the route he had laid out. Kept as a state secret, Ribeiro's map served as the template for the maps used for navigation by all Spanish vessels. Experts now consider it the first truly scientific world map. Still, no map fully evoked the vastness of the ocean until well into the seventeenth century.[37]

Over thirty years later, in 1569, Gerhardus Mercator published his great summary map that represented the three-dimensional globe on a flat surface. To formulate his map Mercator used projections that retained key relationships for navigators while minimizing the distortions of a flattening a curved object. Still used today, his maps played an important role in the subsequent navigations of the world's oceans and marked the passage of the center of mapmaking from Italy to the Netherlands.[38] Mercator dropped Ptolemaic maps as a cartographic foundation, liberating mapmakers to build on information from contemporary explorations.

It takes a powerful exercise of the imagination for us even to conceive of moving from a world of little change to one in which innovations threatened both the powerful and the powerless. Think of the novelties that those living between 1487 and 1527 had to take in: Bartholomeu Dias rounding Africa's Cape of Good Hope; Columbus bumping into islands and continents nobody had ever heard of; Pedro Álvares Cabral accidentally finding the continent of South America on his way to the Indies; Ferdinand Magellan circumnavigating the globe; and Hernán Cortés and Francisco Pizarro conquering magnificent civilizations in Mexico and Peru.

The first reports of New World explorations and exploits found their way into print quickly. Rather than lighting a bonfire, they ignited a fuse that slowly worked its way north, taking a half century to reach England. Oviedo, Las Casas, Columbus, Cortés, Vespucci, Pigafetta, and a few other chroniclers inspired a group of editors and printers to specialize in collecting, illustrating, and publicizing these remarkable ventures into the unknown.

Exploiting the potential knowledge from these voyages of discovery relied on depicting, categorizing, and analyzing the fascinating objects found, a task that took another three centuries. Along the way Europeans laid the foundations for the sciences that have characterized their modernity ever since.

CHAPTER THREE

PUBLISHERS SPREAD THE WORD OF THE NEW WORLD

Drawing of the Tupinamba's ritual greeting from History of a Voyage to the Land of Brazil, *1578, by French explorer and writer Jean de Léry.*

Assessing the intellectual impact of New World novelties on Europeans didn't take place until the middle of the sixteenth century. It took two generations to digest these assaults on their sensibilities. This is not to say that wonder and amazement were not registered immediately, but rather that working out the implications of these surprises through conversations, debates, and writings took time. After the original witnesses had given their reports, a new generation of Europeans grew up accustomed to the reality of an expansive globe. Though the past is often presented as a seamless succession of events, that succession is punctuated by the lived experience of generations. Children can ponder what their parents could only take in.

There were several epistemological problems to deal with. The most serious one was reconciling the existence of these formerly unknown people with the Christian assumption that all beings lived under the judgment of one God. This conviction created a sense of unity and common destiny for humankind. It also undergirded the notion of Providence in which the actual events in time and space were interpreted as messages from God. The workings of Providence demanded that the human recipients of these coded signs acknowledge the communication system itself, which the Amerindians obviously did not. Of course, Christians had long recognized that different groups shared the globe with them, but they had folded them into a coherent narrative about the success and failure of the evangelical work of Jesus.

Now there appeared a kind of diversity that didn't fit in at all. To make matters more confusing, the missionaries going with the Portuguese to the Indies were getting a much clearer picture of

Japanese and Chinese societies. Here were ancient civilizations with sophisticated customs and belief systems, unlike the "naked heathens." This expanded awareness raised the possibility—clearly heretical—that perhaps Christianity was a local and particular religion like those that flourished elsewhere. Philip II of Spain worked closely with the Inquisition, the ecclesiastical body that enforced Catholic orthodoxy, to see that travel accounts were vetted to exclude any material that might give rise to such a heterodox thought.[1] In 1559 the church instituted the Index of Forbidden Books directed principally at Protestant texts but soon encompassing tracts in astronomy that might undermine faith.

The problems didn't end there. Europe itself was becoming more diverse. Paganism was a term of opprobrium sometimes thrown at the humanists for their this-worldly emphasis, but there were more serious manifestations of the Renaissance spirit. In Northern Europe, Martin Luther and others seized upon early Christian texts as avenues to the true Christian practice, stripped of its fifteen hundred years of human accretions. They learned Greek and Hebrew in order to translate the Bible faithfully into vernacular languages. Luther overshot his goal of reforming the church from within and caused a lasting fissure in Christendom between Catholics and the new Protestants, who took Christian doctrine in new directions. Luther's contemporaries, Desiderius Erasmus and Thomas More, were equally inspired by Greek learning, but chose to stay within the Catholic fold.

A long, bloody series of wars, triggered by religious issues but devolving into struggles over territory and power, consumed Europe between the 1580s and 1640s. These conflicts left Protestants and Catholics unwilling to tolerate each other's religion, a hostility that led to national rivalries with countries opting for one or the other form of Christianity. Catholics and Protestants could then effectively control speech and writing within the borders of their countries or city-states.

The breakup of Christianity in the 1530s created new sites of Christian learning less dedicated to the dogma of the old church. When Nicolaus Copernicus, one of the polymaths of the Renaissance, upended conventional knowledge in 1543 with his astronomical proof that the earth revolved around the sun, his *Revolution of the Celestial Spheres* was placed on the Index. Protestants had assumed a decidedly more friendly attitude toward scientific inquiries, though Luther rejected the Copernican hypothesis that the earth circles the sun because Joshua in the Bible had ordered the sun to stand still.

Not only had the various Protestant reformers fractured the unity of Christendom that had prevailed for a millennium, but within Europe fresh identities were being formed around nationalities. Secular languages were officially replacing Latin. As they did, people began to think within their own linguistic tradition.

Probably no other occupational group, aside from the clergy, was more affected by the breakup of Christianity than publishers whose tracts and pamphlets became weapons in the struggle to change minds. Printers operated in more than forty European cities, regularly turning out books, broadsides, and official pronouncements. A clandestine press flourished. Outside their own country, authors found compliant printers willing to publish banned books. Editions of Girolamo Benzoni's hispanophobic *History of the New World*, for instance, delighted readers in Protestant parts of Europe.[2] Smuggling forbidden texts became common. This fierce struggle for the power to define Christian doctrine and practice ultimately enhanced freedom of expression in Europe.

Though initially reticent, Protestants soon came to champion free speech and what we would now call scientific inquiry. Protestants in a kind of logic of opposition encouraged new investigations; their cities became centers of refuge for those who ran into trouble with the Catholic Church, now reinvigorated by the

Spanish Inquisition to look into heretical behavior. Favoring the reading of the Bible, Protestants encouraged literacy, and printers and publishers profited from that extension of literacy. The Netherlands with its flourishing publishing industry was home to the greatest liberties in speech and writing.

Publications about the New World played second fiddle to the avalanche of religious tracts produced by the Reformation of the sixteenth century and the Counter-Reformation of the seventeenth, but they laid the intellectual foundations for the life sciences in Europe. The accounts, which began with Columbus, described and illustrated the unique specimens that scrambled Europeans' mental maps, sending them down a path of inquiry that has never lost its momentum. Religious controversies were old, but questions about the feathers of a penguin, the similarity of llamas to camels, and the way to establish longitude at sea were new and perplexing. They radiated out from practical concerns about diet or navigation to dense discussions of the world of natural phenomena.

The impact of the New World is inseparable from the books and maps that carried information about the discoveries to the private libraries across Europe. Upon his return from the Caribbean, Columbus had composed a letter describing the wonders of the islands he had found on his way to Asia. Within weeks, a printer in Barcelona had published it. He was the first in a succession of publicists ready to exploit the potential news value of the Caribbean explorations.[3]

Print obviously increased the speed and scope for sharing facts and ideas and broke down many of the barriers erected to keep information or opinions secret. The cheap replication of imprints created affinity groups linked by common reading habits. More provocatively, books brought different areas of knowledge into contact, sometimes with disruptive consequences. Over the long run, print encouraged skepticism because it forced readers to con-

front contradictions in rival descriptions of the same phenomenon or even in their own knowledge of events.[4]

Hopes for enlarging their customer base encouraged publishers to market translations in secular languages, moving beyond
the closed circle of those proficient in Latin. Luther's German
translation of the Bible began its long career as the most frequently published book in Christendom in 1534. Printing facil-
itated the preservation of records at the same time that it lured
countless men and women to take up the challenge of writing
about what they knew, observed, and experienced. It even allowed
some erudite authors, like Erasmus, to earn enough from their
publications to free themselves from patrons.[5] The professional
writer had arrived.

Chronicles of explorations, travel accounts, and maps had
a chance to reach educated people throughout Europe because
publishing had become a commercial and cultural phenomenon
of growing importance in the sixteenth century. What appears
remarkable in hindsight is the success of four extraordinary men
who made careers out of publicizing the voyages of discovery.
They were Peter Martyr d'Anghiera, Giovanni Battista Ramusio,
Theodor de Bry, and Richard Hakluyt—two Italians, a Dutchman, and an Englishman. Martyr published his magnum opus, De
Orbo Novo, in 1519, and Hakluyt's The Principall Navigations appeared
in 1589.

All four men drew on the spate of eyewitness accounts that
began with Columbus. Columbus's 1493 letter came out during
the next four years in twenty-two imprints, ten in Latin, five in
Italian, two in Spanish, and one in German. The first collection
of travel literature, Fracanzio da Montalboddo's Paesi novamente ret-
rovati, appeared in Venice in 1507, shortly after Cabral's landfall
in Brazil. Including accounts from Cabral, Columbus, and Vasco
de Gama, it went through fifteen editions in four languages.[6]
These four men produced dozens of attractive, readable volumes
with firsthand accounts from the navigators themselves.[7]

By the time of Cortés's conquest of Mexico fourteen years later, another hundred publications about the New World had appeared in Latin as well as every major language of Western Europe except English. Animated by very different motives, Ramusio, Martyr, de Bry, and Hakluyt collectively sustained and amplified the interest in European voyages of discovery. Martyr was an Italian scholar who ended up at the Spanish court at an early age. An erudite Venetian diplomat, Ramusio was drawn to the discoveries by his love of geography. His *Navigations and Travels*, the first volume appearing in 1550, expanded on Martyr's collection about them. His scientific interests distinguished his texts from the more romantic versions that Martyr favored. Hakluyt gathered many of both authors' publications along with new material on discoveries to nudge England toward colonization in the Western Hemisphere. He infected de Bry, after a distinguished career as a publisher in Frankfurt, with his cherished goal to plant Protestant settlements in the New World.

Martyr became the first publicist of the voyages of discovery through a somewhat circuitous route. He must have been an exceptional scholar and a well-connected courtier, for when he went to Rome at the age of twenty in 1477, his career took off. He met a number of significant figures—ecclesiastical dignitaries, courtiers, generals, and diplomats, including the Spanish ambassador. Ten years later Martyr left with the ambassador, who was returning to Spain, where Martyr turned soldier and joined the attacks on the Muslims holding on to Granada. Martyr thrilled to the exploits of the Spanish, intent on expelling the Muslims who had been there since the seventh century. He compared the *Reconquista* to the Trojan War and Caesar's campaigns in Gaul.[8]

Returning to court after the war, Martyr took holy orders and became chaplain to King Ferdinand and Queen Isabella. Columbus's visit to Barcelona to report to his sovereign sponsors on his 1492 crossing of the Atlantic—the court appearance that Las Casas also witnessed—convinced Martyr that he should stay in Spain. A generation older than Oviedo and Las Casas, Martyr secured a monopoly on information about Spain in the New World.

His residence in the Spanish court gave Martyr access to the best records available of the first explorations. He remained to become the chronicler of the Council of the Indies that Ferdinand and Isabella's successor, Charles I, formed. Martyr published mainly for readers in Italy and Southern Europe. Ramusio too wrote for Italian and Spanish Catholics. Hakluyt confined himself to the British, while de Bry strove to reach the Protestant public in the Netherlands and Germany. As the colonizing impulse shifted from southern to northern Europe, so did the center of publishing.

In the sixteenth century publishers, though always businessmen and often craftsmen as well, hobnobbed with humanist scholars as well as statesmen. Their intimate knowledge of original texts gave them a valuable expertise that others sought out. Offering eyewitness narratives, their publications became the focus of discussions in council chambers, towns, universities, and monasteries throughout Europe.

An interest in explorations, nurtured by their close cousins— geography, ethnography, botany, and zoology—acted like a magnet for new associations. Common interests cemented personal relationships, often maintained through correspondence. Books and letters linked enthusiasts across national borders. Like the humanists, bound by a love of ancient texts, those following the voyages of discovery introduced something new to European society: affinity groups. These avid readers formed a Republic of Letters in a world of monarchies.

Swept up in Martyr's net were manuscripts on Columbus's four voyages, fragmentary accounts from explorers who followed in his wake, reports to the Council of the Indies, miscellaneous letters, information on Balboa's discovery of the Pacific Ocean, as well as the personal narratives that explorers, returning to court, dictated to Martyr. He also conducted personal interviews. Both John Cabot, who explored the North American coast for the English, and Amerigo Vespucci visited Martyr, sharing maps and stories. Martyr saved the ephemera as well as the essential documents of the conquistadors and preserved them in his exquisite Latin prose. These formed the raw data of conquest—firsthand and fresh—as the Spanish imposed their rule over the islands of the Caribbean and the northern part of South America.

Although Martyr had access to official records, there was nothing official-sounding about his account. He assumed the cozy yet confident voice of a storyteller who will now regale his audience with marvelous tales of ardor and piety. He begins his magnum opus by referring to "a certain Christopher Columbus, a man from Genoa." Columbus could not have had a more devoted admirer, and he needed one, for he had to struggle within the cauldron of royal politics to secure his awards. His fledgling settlement on Hispaniola was rife with intrigue and treachery, which Columbus attributed, according to Martyr, to the fact that the Spaniards were "more addicted to sleep and idleness than work, more keen on mutiny and revolts than peace and tranquility."⁹ Martyr became a partisan of Columbus, to whose doings he gave exquisite attention. He called his volumes "decades." The first one, which was read in manuscript for ten years before publication in 1511, probably served as a lobbying piece for Columbus's heirs.

A diplomatic mission in 1510 carried Martyr to Cairo, where he visited the ruins of Alexandria, the pyramids, and the Sphinx. His account of this journey reintroduced his readers to the ancient Egyptians. When Charles I formed the Council of the Indies in 1520, he named Martyr its chronicler, thus enhancing his career as a Gabriel summoning Europeans to news about the larger world around them. His fifth and eighth volumes featured the conquest of Mexico with its startling descriptions of Aztec cities. Others had a miscellaneous quality covering both Spanish exploits and descriptions of Amerindian mores. In 1530, four years after his death, all nine of Martyr's decades were published together in *De Orbe Novo*, the first complete history of the New World.

Martyr was a very competent geographer and a superb critic and editor of the many texts he handled. When Juan Ponce de León encountered the Gulf Stream in 1511, Martyr realized that it would hasten the return of Spanish ships across the Atlantic. What is more important, Martyr saw that the discoveries opened up a vast new canvas for historians. The detailing of dynasties, rivalries, and warfare would have to move over to make room for the stirring narratives of global explorations that began in 1492.

Martyr's decision to highlight descriptions of the people whom the Spanish encountered in the New World was a shrewd response to what were surely the most thought-provoking of all things reported back to Europe. Columbus had commissioned a friar working with natives on Hispaniola to prepare a report on their ways, and Martyr drew on it.[10] He had also listened when Las Casas urged the Spanish king to rein in the violent suppression of the Indians. He became attentive thereafter to reports of Spanish abuses. He clearly saw the contradiction between exploiting native labor to extract wealth and nurturing the people to secure their conversion to Christianity. Martyr's opinions of the native men and women remained ambivalent.

The benign climate, naked inhabitants, and luxuriant flora of the Caribbean islands brought to many minds the image of a lost paradise. Columbus, after all, had compared the mouth of the Orinoco in Venezuela to the Garden of Eden. Yet for Martyr, the Indians' nakedness signaled not a different culture but a lack of culture, an attitude understandable given the complex meanings associated with dress in European courts. At the same time he characterized the pre-Columbian years as a Golden Age, despite his Eurocentric assumptions about Amerindian cosmology. He devoted the last book of his first decade to the religious beliefs of the Taino Indians, a subject that absorbed him for many years. Finally he concluded that the natives had blank minds upon which could be imprinted the languages and knowledge of European civilization.

Although not a printer, Martyr cared about the physical composition of his works. He favored the new roman typeface patterned on the characters of antique Roman inscriptions, even though his Spanish patrons preferred the old Gothic fonts.[11]

Martyr's twenty-nine-year monopoly on publishing material about the Spanish Indies came to an end when Hernán Cortés published his first letter in 1522. Here at last was a conquistador talented enough to tell his own story. Cortés followed his first lively *Letter from Mexico* with three more in 1522, 1523, and 1525, succeeded in the next decade by firsthand accounts of a different kind. A fine stylist with a critical intelligence, Cortés did justice to the Aztec civilization through the care with which he detailed its accomplishments and ultimate conquest.[12]

After his death in 1529, Martyr's friends arranged to have much of his correspondence published. The 812 letters in Martyr's *Opus Epistolarum* became a great archive for information about the reception at court of Spain's navigational triumphs during the momentous years between 1487 and 1525. In that span of years, Martyr had single-handedly gathered the records from the dozens of expeditions that radiated out from the Spanish foothold in Hispaniola.

Gonzalo Fernández de Oviedo published the first part of his great *Natural History of the Indies* the year Martyr died. Readily recognized for its comprehensiveness and objectivity, this publication heralded the arrival of the Americas as a serious subject for study, not just the location of European military triumphs. The work of Bartolomé de Las Casas, the other "Herodotus" of the Spanish Indies, appeared in 1542.

Martyr's successor in the grand enterprise of publishing documents on the voyages of discovery was Ramusio, who started
collecting explorers' firsthand accounts in the 1520s. Travels into
Central Asia and China also attracted Ramusio's attention. He
was responsible for sifting through some 150 manuscript versions
of Marco Polo's thirteenth-century trips. So popular was Polo
that many, many copies of his travels had been scattered through
Europe's courts and universities. As with the children's game of
telephone, copies of copies begat dozens of variations of Polo's
stories. Ramusio produced a text that scholars have relied upon
ever since.[13]

As befitted the scion of an aristocratic family of Venice, Ramusio became a secretary to the Venetian ambassador to France
when he was twenty. His career in the republic's diplomatic service gave him access to the stream of news about the voyages to
the New World and elsewhere. Fluent in several languages and
ardent about geography, he arranged for translations of the writings of Vespucci and Pigafetta. Responding to his generation's
transition to reading in secular languages, Ramusio translated his
material into Italian, the most popular language of the early sixteenth century.

Ramusio published the first account of Cabot's 1497 voyage to
Newfoundland for the English and a decade later that of Jacques
Cartier's expedition for the French, in which Cartier found the
estuary of the St. Lawrence River and named it Montreal. Ramusio's concentration on firsthand accounts from the explorers
resonated with those who shared his fascination with how the
boundaries of geographic knowledge were being steadily pushed
outward. His bringing together all of the known sea routes and
peoples Europeans had visited brought the whole panorama of
explorations to the reading public.[14]

Described as a "treasure trove of travel accounts," *Navigations
and Voyages* became the model for subsequent publications once

it became apparent that there was a ready market for them. For Ramusio, the dedicated humanist, there was a more serious purpose than mere sales. He wanted his readers to engage with the new knowledge and sort out the errors they had imbibed from ancient authorities like the first-century philosophers Ptolemy and Pliny. In his preface Ramusio explicitly said that he had brought this material out for "scholarly people" as well as "lords and princes."[15] He wrote as an authority, whereas Martyr had assumed the persona of the insider who is privy to all the particulars of an astonishing succession of events.

In England, far from the center of explorations and their publicists, Richard Hakluyt gave himself the mission of awakening his fellow Englishmen to the promise of riches and glory held out by prospective settlements in North America. It was about time. Most of the public there remained unaware that Columbus had sought support through his brother from King Henry VII or that Cabot had explored Greenland, Newfoundland, and the east coast of North America under the English flag in 1497 and 1498.

At age thirty, in 1582, Hakluyt accompanied the English ambassador to the French court as chaplain and secretary. He spent five years in Paris, where he learned to his chagrin that among the Italians, Spanish, and Portuguese the English were mocked for "their sluggish security, and continual neglect" of venturing across the Atlantic.[16] That indifference was short-lived. In the last decade of the sixteenth century, English adventurers like Walter Raleigh, Thomas Cavendish, Francis Drake, and John Hawkins took to the Atlantic sea-lanes with great gusto. Their derring-do on the high seas added to the discomfiture of Spain, already angered by England's break with the Catholic Church. Another daredevil, Martin Frobisher, was plunging into the icy realms of Canada to find a northwest passage to the Orient when he wasn't harassing the French.

Hakluyt's first publication, in 1582, was a somewhat motley

arrangement of dozens of narratives and documents. Its wordy title conveys his focus: *Divers Voyages Touching the Discoverie of America ... Made First of all by our Englishmen and Afterwards by the Frenchmen and Britons.* Still, the book earned him a reputation among people in high places. Next Hakluyt prepared an English translation of a Latin account of Cabot's voyages that established England's claims to North America. Hakluyt's major work, *Principal Navigations, Voiages, and Discoveries of the English Nation,* appeared in 1589, when the menace of the Spanish Armada no longer hung over the nation. Nothing this comprehensive about exploration had ever been published in English. Divided into three parts, it covered explorations to Africa, the eastern Mediterranean, and to the northeast, with the longest part devoted to North America.

Drake and Hawkins had been menacing the Spanish silver fleets that returned each year with riches from the king's Mexican mines. Cavendish followed Drake in circumnavigating the globe. Elizabeth knighted them both, even though they often continued their marauding after peace treaties had been signed. Ready to move beyond piracy, Raleigh asked Hakluyt to write a proposal for planting a colony in North America; Hakluyt personally presented it to Queen Elizabeth.

Hakluyt pulled out every stop in his appeal. English settlements could turn the improvident poor into commodity-producing farmers; English Protestantism needed to be planted in the New World; transatlantic trade would nurture a strong navy and a fresh cadre of experienced seamen. Hakluyt wanted England to focus its ambitions in the New World, but he lent his support to the founding of the English East India Company in 1600 after Drake circumnavigated the globe and Cavendish sailed English ships into the Indian Ocean. Whereas other printers and publishers turned out books to appeal to a general readership, Hakluyt sought the attention of royal advisers, aristocratic adventurers, and English overseas merchants whom he hoped to prod into action.

Raleigh, Queen Elizabeth's most dashing courtier, focused on the area between Venezuela and Brazil after reading a Spanish account of a golden city at the headwaters of the Caroni River. Ever the promoter, he dashed off a prospectus describing the lush beauty of the Guyanas. In 1595 he sailed to the area himself. The following year he wrote up his adventure, dazzling his readers with details of gold mines that he had never seen. The El Dorado legend persisted, although no mines were ever found there, but the region did contain turbulent rivers and a wild terrain that went from jungle to savannah. His contemporary, Frobisher, infected by the same passion for gold, brought back tons of what turned out to be iron pyrite from Baffin Island. Even this fiasco failed to stifle English dreams of landing in a gold mine as the Spanish had.[17]

Casting his eyes on the long Atlantic coastlines that stretched between present-day North Carolina and Maine, Raleigh next outfitted several colonial ventures. He gave the name Virginia to the area north of Florida in honor of his sovereign Elizabeth, the virgin queen. Ever on the outlook for novelties, he introduced tobacco and the potato to England as well as to Ireland, where he had estates. At the same time the French, also late to enter the game of colonization, fixed their eyes on Florida. The Spanish crown considered both areas theirs, but lacked the means of expelling the intruders.

In a dedication to Raleigh in his English edition of Martyr's *De Orbe Novo*, Hakluyt wrote that "geography is the eye of history." A somewhat enigmatic thought, it could be interpreted as saying that studying geography would lead to a different kind of history for England. Lavishing praise on Martyr, Hakluyt said that he accomplished for the Spaniards what Homer had done for Achilles, perhaps imagining that he might do the same for Raleigh.[18] But, alas, when James I followed Elizabeth to the English throne, Raleigh lost favor at court. Worse, he was tried and convicted of treason for plotting against the king, who suspended his death sentence and imprisoned him instead.

The thirteen years Raleigh spent in the Tower of London gave him the leisure to write his *History of the World* and to father a son. James released Raleigh in 1616 to make one last voyage to the Caribbean, where again he failed to find gold. Not wishing to go home empty-handed, his men attacked the Spanish settlement on the Orinoco River. Outraged, Spain, now at peace with England, complained to King James through its ambassador. The king had Raleigh executed forthwith when he returned home.

Almost as unfortunate was Anthony Knivet, who had accompanied Cavendish on a second voyage in 1591. Expecting to reap the riches of the Orient, Knivet instead became stranded in Brazil as the slave of a Portuguese sugar mill operator. Passed from owner to owner, Knivet finally became a trader with the cannibalistic Tupi-speaking Indians who stayed their appetite for Knivet in order to bargain with him for metal tools. After several failed escape attempts followed by reenslavements, Knivet eventually made it back to England, where Hakluyt published his memoir in 1625. It both horrified and thrilled the English public at the very time that they were seriously engaged in establishing a colony at Jamestown.[19]

The forming of the Virginia Company signaled the recognition that it took deep pockets to found a colony. Even then it was fraught with perils. Fittingly, Hakluyt was named cleric of the Virginia colony, though he remained homebound in England until his death in 1616. Still, Hakluyt could take great pride in knowing that he had succeeded in getting his countrymen committed to establishing a foothold in that part of the New World they could legitimately claim.

Convictions about a common human destiny under God died slowly. When English Puritans established colonies in North America two decades after the Virginia settlement, their announced purpose was to set an example that would persuade Catholics and Protestants to unite in emulating their Christian practices. John Winthrop's "city upon a hill," so often quoted about the United States, was evoked in the hope that Catholics would adopt Protestant reforms once they saw them in a pure form in England's Massachusetts Bay colony.

Hakluyt met de Bry in London, although de Bry had settled in Frankfurt after living in Liège, Strasbourg, Antwerp, and London. He kept moving in part to avoid persecution for his strong Protestant convictions. That in turn explains his attraction to firsthand accounts of the explorations of North America's east coast, where the hated Spanish Catholics had left only a small mark in Florida. In choosing to publish material on North American expeditions, de Bry reflected and nurtured an already strong German interest in the discoveries. German financiers and mining engineers had become early participants in the European scramble for riches in the Western Hemisphere. A German immigrant printed the first book in Mexico in 1539.

German miners sent under contract from a Saxony firm to Santo Domingo returned with stories that helped shake Europeans out of the fantasy that the indigenous people of South America somehow resembled Greek statuary. Two German soldiers helped in this. Captured by Portuguese officials in Brazil, they managed to free themselves and get back home. One of them, Hans Staden, wrote a tale whose title, *The True History and Description of a Country of Wild, Naked, Cruel, Man-eating People in the New World of America*, epitomizes the lurid testimony in it.

Staden's book became an international bestseller when de Bry published it 1557, followed by seventy-six editions of translations into the major European languages. Staden described being ambushed, taken naked back to the Tupinamba settlement, and forced to shout out, "Here I come, food for you." He designed fifty graphic woodcuts to illustrate his salacious tale of a ten-month captivity among cannibals. He recounted how the Indians had taken down a cross that he had erected. When they became inundated with rain, they pleaded with him to get his god to stop the rain. Readers doted on such anecdotes. They were reassuring evidence that their god was the true one and Amerindians possessed no magic that could threaten them.[20] This incident apparently saved Staden from becoming his hosts' dinner.

Jean de Léry, a French Huguenot, left a more serious account of the Tupinamba Indians gleaned from an earlier expedition made by Protestants reconnoitering the land for a possible settlement site. In his *History of a Voyage to the Land of Brazil*, he explained that he had joined a group of fellow Protestants in their new colony on an island in the Bay of Rio de Janeiro "out of curiosity to see this New World." When the colony moved to the mainland, he came into close contact with man-eating Tupinambas. Living close to them for over six months, he could furnish a detailed and somewhat grisly account of how they prepared human bodies for food. Between his trip and the publication of his history in 1578, de Léry witnessed one of the biggest atrocities in the ongoing conflict between French Protestants and Catholics. Protestants thronged to Paris to celebrate the wedding of one of their heroes on St. Bartholomew's Day in 1572. Catholics hunted down the soberly dressed Protestants in the streets, and killed thousands of them with cutlasses, knives, bats, bricks, pipes—anything that came to hand. For de Léry, the Massacre of St. Bartholomew's Day, as it became known, put the cruelty of cannibalism in the context of man's inhumanity to man rather than indicating a hideous flaw of pagans.[21]

The brutalities of the St. Bartholomew's Day Massacre outraged the famous essayist Michel de Montaigne who, like de Léry, saw it as a mockery of the common assignment of barbarism to Amerindians. For him the presumed naturalness of the people of the New World became a reproof to Europe's excessively civilized, yet vicious, societies. Montaigne learned about the people of the New World by talking to some natives that his king had brought back to Rouen. He also had long conversations with his servant, who had participated, as a seaman, in a French voyage to Brazil, where he remained for a decade. In his celebrated essay "On Cannibalism," published in 1580, Montaigne makes much of the truthfulness of simple people like his servant who can be believed because he lacks the wit to embellish his tale.[22]

Montaigne's essay was a paean to tolerance. His argument: we castigate in others what are merely ways different than ours. "I think there is more barbarity in eating a man alive than in eating him dead; and in tearing by tortures and the rack a body still full of feeling ... having him bitten and mangled by dogs and swine ... than in roasting and eating him after he is dead."[23] Homogenizing all the diverse tribes of the New World into one people, Montaigne contrasted his fellow Europeans with natives who had "no kind of commerce, no knowledge of letters, no science of numbers, no title of magistrate or of political superior, no habit of service, riches or property, no contracts, no inheritance, no divisions of property, only leisurely occupations...." He described them as given to dancing, hunting, and drinking. He even extolled the valor that led the tribes of Brazil to eat their prisoners of war.[24]

Shakespeare joined Montaigne in making invidious comparisons between Indians and Europeans even though he recognized that the Indian utopia Caliban described in *The Tempest* expressed more of a yearning than a fact. The discoveries' capacity to promote a sophisticated relativism had just begun. The publishers of accounts of expeditions and forays to establish European colonies in the Western Hemisphere created and sustained an interest in these eyewitness accounts.

Although he was twenty-four years younger than de Bry, Hakluyt managed to talk the noted German publisher into promoting Protestant colonization. Late in his life de Bry picked up Hakluyt's baton and started editing and printing works on Protestant New World settlements. He was lucky in having talented sons who could complete his new project and fortunate in possessing what Hakluyt lacked as a publicist: a printing firm with skilled engravers who could make handsome illustrations to adorn his volumes.

Perhaps even more important, de Bry came into possession of drawings from two gifted artists: John White, who had accompanied the English colonizers to North Carolina and Jacques Le Moyne, whom the French king had sent on an expedition to Florida. De Bry had an eye for art. As a bookseller he could see the appeal of these compelling depictions of the unfamiliar people, animals, and plant life that filled the newfound lands. With the manuscripts and published texts, watercolors, and gouaches (a thickened type of watercolor) that he collected in London, he was ready to begin the project that would give him lasting fame: his *Grand Voyages.*

What made de Bry's first two publications of immediate and enduring value were White and Le Moyne's illustrations from which de Bry's staff made copperplates, a technique unavailable to Hakluyt. These stirred the imagination of thousands of readers who could now imagine the New World through evocative pictures.[25] The semiannual book fair in Frankfurt attracted printers and publishers from all over Europe and a significant portion of their clients as well. The publishers' catalogues became bibliographies for those eager to pursue a special topic like seaborne exploration.[26]

The first work in de Bry's great travel series came out in 1590, when de Bry was over sixty. It featured Thomas Harriot's *A Briefe and True Report of the New Found Land of Virginia*. Harriot, an English mathematician who had tutored Raleigh and other sea captains in navigation, first sailed with the fleet to the North Carolina coast in 1584, accompanied by White. The two men were charged with exploring the land and recording in words and pictures what they saw. White had already been in the Western Hemisphere, traveling with Frobisher in 1576 to the Hudson Bay area where he had sketched a lively scene of an Eskimo attack on the intruders.

Harriot, who learned the Algonkian language, made careful observations and mapped the area while White produced handsome drawings of the Algonkian people at play, at work, in repose, and during religious dances. Despite this promising beginning, the first contingent of settlers, feeling neglected, jumped at the chance to return to England when Drake happened upon them as he was sailing up the Atlantic coast.

Three years later, in 1587, White led another set of colonists back to Roanoke, among them his pregnant daughter. Forced to repair to England for more provisions, he was blocked from returning to the settlement because all English ships had been commandeered to fight the Spanish Armada. When he did get back, three years later, no trace of the colony remained. "The lost colony of Roanoke" found a permanent place in American history in part because of the legend of the first English person born in America, White's granddaughter, Virginia Dare.

Unlike Hakluyt's spare English edition of *A Briefe and True Report of the New Found Land of Virginia*, de Bry's included engravings of White's very fine watercolors. The popularity of this first volume on New World explorations convinced him to continue with the project. An astute judge of the diverse audiences he sold to, de Bry modified his texts as they moved from Latin publications to translations into the major vernacular languages of Western Europe. Engravings ran up the cost of volumes, so individuals, or the learned societies that often sponsored these publications, had to pay for them through advance subscriptions.

De Bry also brought back from London a cache of handsome botanical and ethnographic drawings that Jacques Le Moyne had done on an ill-fated French colonial expedition. The French king had sent Le Moyne along as official recorder for this would-be refuge for French Protestants in northern Florida, which the Spanish obliterated. Le Moyne was one of a handful of survivors of that disastrous venture. De Bry actually wrote most of *The Short History of Those Things that Befell the French in the American Province of Florida* in order to provide a text for Le Moyne's forty-two illustrations of scenes and maps. Returning to Europe, Le Moyne found a patron in Raleigh and remained in England as an illustrator of flowers for a readership that now demanded botanical specificity.

De Bry's zeal as a Protestant led him to juxtapose illustrations of statuesque nude natives in classical repose alongside Spanish conquistadors dressed for battle. He brought out a new Latin edition of Las Casas's exposure of Spanish abuses with particularly shocking illustrations of Spanish soldiers disemboweling pregnant Indian women. These images found their way into successive editions in Dutch, German, French, and English.

Lest anyone miss the point, de Bry entitled his edition *The Mirror of Spanish Tyranny in Which the Murderous, Scandalous and Horrible Deeds are Recorded Which the Spaniards Have Committed in the Indies.* He also did something intellectually provocative. He included depictions of the ancient Picts, who lived a millennium earlier in England. This suggested to many readers that the Amerindians might best be thought of as being like the earlier inhabitants of Great Britain—an idea powerful enough to provoke the interpretation of human origins as evolving over time.[27]

De Bry reissued Staden's account of his Brazilian captivity in the third volume of his series. Woodcuts of Staden's crude drawings of the wild, man-eating Tupinamba made his story particularly compelling, as did other, more realistic depictions of native peoples in De Bry's publications.[28] Although the Portuguese were the first to establish a colony in Brazil, they left it to others—Frenchmen and Germans—to write about the land. They were a notably reticent group, Ferdinand Magellan being an exception

Theodor de Bry's engraving of "A True Pict" from A Briefe and
True Report of the New Found Land of Virginia, *1588,
suggesting that the original inhabitants of Great Britain were
much like the native people of the New World.*

in getting Antonio Pigafetta to record what he knew would be a
historic voyage.

Other volumes in de Bry's series contained the heroic depiction of Pocahontas's dramatic rescue of Captain John Smith as
well as scenes of the grisly murders in the 1622 Indian massacre

of Jamestown settlers. Both became seared in the historic memory of the English and their Virginia colonists. In 1634, thirty-six years after de Bry's death, his sons completed the twenty-five volumes of the *Grand Voyages*.

The publications of Martyr, Ramusio, Hakluyt, and de Bry of firsthand accounts of the voyages of discovery were almost as significant as the voyages themselves. Certainly Hakluyt thought that they were, toasting the "men of powerful intellect who have sealed the record of their glorious deeds in monuments of literature that will last forever."[29] The publicists transformed these expeditions, authorized by monarchs and funded by merchants and bankers, into a cultural possession for all Europeans.

Reprinted throughout the next two centuries, these volumes replaced the ephemeral memory of living participants with the permanence of texts. The volumes traveled widely, finding readers wherever there were roads. Reading about the New World begat conversations, and conversations generated disputes. Disputes raised questions, keeping alive the memory of the hundreds of novelties brought back from the voyages of discovery. At the same time, the Protestant presses in England and Holland were providing the principal vehicles for the writings of naturalists whose interest had been aroused by the astounding contents of the New World. The printing press became the vehicle for popular education, particularly about life outside of Europe.

The existence across the Atlantic of strange people with truly weird mores shook Christian confidence in the universality of their religion, which until now had only been marred by the existence of stubborn Jews and heretical Muslims. With travelers going to Japan, China, the Indies, the Middle East, and North Africa, it dawned on many thinkers that there were too many different people to fit into Noah's Ark, too much empirical data to dismiss as the tales of "roving liars."[30]

Alongside the Christian understanding of different peoples

grew a discourse about civilizations. Capacious enough to deal with the steady intake of new information, it spawned efforts to explain how diverse groups of men and women had fashioned norms, along with elaborate explanations of their origins. Thinkers began to correlate patterns of behavior with climate, topography, and isolation.

When the first titles in the literature of discovery appeared, many an old woodcut of the Garden of Eden had been freshened up a bit to do double service as a depiction of the New World. More than pragmatism was involved. Europe's intellectual traditions had always pointed backward to golden eras, either that of the Garden of Eden or of ancient Greece. This perspective assumed that human societies had been degrading for centuries. For Christians only an act of God could save them.

Slowly a counter-idea emerged. Perhaps de Bry's hint was correct, and the Amerindians resembled Europeans of a few centuries earlier. If so, might not the present represent a general improvement in the level of human accomplishment? It took a long time for this radical idea to gain acceptance, but when it did, it completely reversed Europeans' time orientation. Then it became the future that contained hope for humankind and not the so-called Golden Ages in the past.

One of Michel de Montaigne's contemporaries, Louis Le Roy, captured this new spirit in his *De la vicissitude en variété des choses en l'universe*. Reveling in the fact that the entire human race and its habitats were now known, Le Roy gave credit for this European accomplishment to the printing press, the mariner's compass, and gunpowder. His is one of the first of many efforts—both humble and arrogant—to offer an explanation for why Europeans could do what other people couldn't seem to do: travel widely and impose their will wherever they went.

Narratives of exploration were so widely circulated that they finally found their satirist in 1726 with Jonathan Swift's *Travels*

into Several Remote Nations of the World, in Four Parts, By Lemuel Gulli-
ver. Gulliver's travels took him to Lilliput, just north of Tasmania, and Brobdingnag, somewhere east of Japan. Swift had a field
day mocking the weird names littering most travel books, to the
amusement of generations of readers. In addition to the Lillipu-
tians and Brobdingnagians were the Laputans and Houyhnhnms
whom Dr. Gulliver encountered on his always-disastrous trips.
Even those who haven't read *Gulliver's Travels* recognize the yahoo
as an ignorant bumpkin and muckrakers in the Laputans' practice
of exposing conspiracies by raking through the muck of untidy
possessions of suspicious persons.

The novelties of the New World provoked new habits of
thought in the minds of those liberated from the cages of con-
vention, as Swift wittily pointed out. Men and women became
intrigued not just by the unexpected diversity of things in the
Western Hemisphere. Through contrast, aspects of their own
world now became salient. Where before the intellectual elite had
concentrated upon Hebrew texts or Greek epigraphs, increasingly
there emerged experts on frog mating habits and the stamen of
flowering plants. Such inquiries drew on a different set of talents than those of the linguist and scholar. New questions were
reorienting European culture toward the natural phenomena all
around them.

❊ CHAPTER FOUR ❊

COLLECTORS, MENAGERIES,
AND NATURALISTS

The Brazilian Ball for Henry II in Rouen, October 1, 1550,
painted by an anonymous artist.

In the two centuries after Columbus's voyages, Europeans became increasingly engaged in the world around them. Excited by the discoveries in the Caribbean and Pacific and influenced by the inquisitiveness of the ancient philosophers, they launched hundreds of inquiries about natural phenomena. While doing so, they ran athwart the Catholic Church's teaching that curiosity drew people to the carnal world and away from things spiritual. Christianity claimed to be a universal religion, its mandate to win converts imperative. Even after the Reformation, both Catholics and Protestants instilled their adherents with evangelical zeal, encouraging them to reach out and bring heathens into the fold. This meant that the whole world offered a field for conversion. When Tunisian traders on the Malabar Coast asked Vasco da Gama's sailors what had brought them so far, they replied, "Christians and spices."[1] This evangelical impulse thrust Europeans into strange and intriguing new places. The emerging fields of botany, zoology, geology, and anthropology, prompted by the discoveries outside Europe, depended upon this global travel.

New instruments arrived to play a part in the aroused inquisitiveness of Europeans. A century after the Spaniards started exploring the lands washed by the Caribbean, an optical expert in the Netherlands figured out how to manipulate convex and concave lens in such a way as to refract rays of light for magnification. By 1609 Thomas Harriot working outside of London and Galileo Galilei in Venice had observed the moon and depicted it in drawings. Galileo was the more skilled as an astronomer and, as befits a native of Florence, the better watercolorist. His keen sense of competition drove him to search the heavens with his spyglass, as telescopes were first called.

As with Columbus and Magellan, visionaries need patrons. Galileo had one in Grand Duke Cosimo II of Tuscany. When Galileo located the four largest moons circling Jupiter, he named them the Medician stars to honor Cosimo and his three brothers. These dependably circling moons became a navigational aid. Galileo's observations of Jupiter convinced him that Copernicus was right: the earth did spin around the sun. His *Starry Messenger* of 1610 describing Jupiter, the moon, and the stars in the Milky Way revolutionized thinking about the planetary system.

The spread and perfection of the telescope turned astronomy into a science of observation and calculation. The century that began with Galileo and ended with Isaac Newton laid the foundation for modern physical science.[2] Galileo and his mathematical peers actually saw themselves as voyagers, using the imagery of exploration to explain their exploits as astronomers. They spoke metaphorically of sailing into the unknown, avoiding shoals, and finding gems of knowledge at the end of their quests.[3] Galileo's admirers extended the comparison by calling him "Florence's second Amerigo."

Prior to this period, in a way that is difficult for us to comprehend, medieval scholars had thought of nature principally as a source of signs carrying messages or truths. They had believed that the curves, angles, and corners found in nature betokened some specific information about the nature of the universe—and that the universe was a moral one. Allegories, resemblances, and symbols, thought to convey eternal verities, drew these scholars into texts that they decoded rather than investigating the natural objects that inspired them. Christian writers thought in terms of the concept of Providence by which God's intentions were communicated through natural events. A shipwreck could indicate punishment for the pride of the owner or the debauchery of the sailors; comets might predict an attack or impending war.

Animals, like the anthropomorphic characters in Aesop's fables, figured in morality plays. Ascribed to a Greek slave of the mid-fifth century BCE, the famous stories actually were an accumulation of Indian, Sumerian, and Egyptian tales. Well into the modern era, children read about the tortoise's race with the hare and the frustrated fox who disparaged as sour the bunch of grapes he couldn't reach. In medieval emblem books, the habits of animals were detailed for the edification of readers. Emblems with their stylized depictions of animals, closely associated with heraldry, conveyed homilies involving turtles, hares, foxes, and others. Into this world, cordoned off from reality by their pedagogical functions, came objects from the New World so arresting that they diverted attention from what they represented to what they actually might be.

Like Aquinas with Aristotle, Christians had to meld their new experiences with the precedent-shattering voyages of the fifteenth century into their old dogma in order to answer concerns of potential converts as well as doubts that they themselves might harbor. José de Acosta, a renowned Jesuit theologian who spent fifteen years at the end of the sixteenth century in Latin America, took up the challenge. He called "the natural curiosity of men" a handmaiden to God's desire that the light of the Gospel should shine on those still living "in the darkness of their errors." As if to prove his own credentials as an inquisitive man, Acosta studied the Incan and Aztec calendars and the complex meanings in their hieroglyphs. Still, as late as the eighteenth century, some clerics persisted in denouncing inquisitiveness as a "spiritual adultery of the soul," but it was a losing cause.[4]

Fascination with the objects Europeans brought back from their travels prompted people to start talking about "curiosities." Soon they were building cabinets of curiosities: display cases of intricately patterned butterflies, memorable leaves, and arresting animal fossils. European noblemen always strove to demonstrate magnificence in their person, their court, and their possessions. As royal bureaucracies grew so did the number of officials with princely incomes. This wealth and the increasing number of enticing objects to possess fueled a competition among collectors. Demand met supply. To the familiar acquisition of paintings, books, coins, and medallions could now be added Indian head-dresses, corals, shells, toucan beaks, tobacco pipes, Chinese porcelain, Turkish shoes, elephants' heads, and even preserved native bodies. Sometimes the cabinets took over more and more rooms in a house, occasionally overtaking the house itself, laying the foundation for a public museum.

The indigenous people of the New World had long attracted royal attention. The Aztec athletes whom Cortés brought back to Spain played a very ancient kind of soccer, called Ulama, before the court at Seville. More amazing than their prowess at the sport were the rubber balls they used, for Europeans had only hollow balls made of leather. Having witnessed such games in the New World, Oviedo drafted a memorable description of them, explaining that "even if they are only let slip from the hand to the ground, they rise much further than they started, and they make a jump, and then another and another, and many more, decreasing in height by itself, like hollow balls but more so." The Amerindians had been extracting latex from rubber trees for hundreds of years, perfecting a product that could be used to waterproof clothing as well as create bouncing balls.[5]

Henry II of France also entertained his court in Rouen with ceremonies performed in a reconstructed Indian village. There in 1550 the Tupinamba, who had held Hans Staden captive, staged fights with their enemies, the Tabagerres. Both groups played their parts with such fierceness that they scared onlookers with the ruckus. They were later shipped home, but not before Montaigne had familiarized himself with them long enough to compose his famous essay on cannibalism. Such pageantry continued to appeal to European spectators. In 1577 Frobisher arranged duck shoots with kayaking Eskimos in Bristol harbor.[6] Still, engagement with the wonders of the Western Hemisphere did not prevent impecunious collectors from melting down elaborate gold Incan decorations when they were hard up, as did Charles I.

Inspired by classical texts, humanists developed a taste for what we would call antiquities—medals, coins, and tombstone inscriptions of the Roman era. Many of these had literally lain around Italy and southern France in splendid neglect. With a kind of intellectual alchemy, relics that were once considered rubbish became precious collectibles. As traveling became more popular in the seventeenth century, going to see collections open to visitors became an integral part of travelers' itineraries. One enthusiast covering the breadth of Western Europe located 968 collections of antiquities alone. The ideal museumgoer, a contemporary explained, was a man of "judicious curiosity" rather than an unbridled appetite for wonder."[7]

Princes and prelates emulated Solomon and Alexander the Great by putting together menageries to delight their visitors and demonstrate the reach of their dominion. A few of these collections were open to the public; others required a letter of introduction before entry could be gained. Bestiaries, herbariums, and lapidary exhibits abounded. Cardinal Ippolito Medici kept what he called "a troop of barbarians" who spoke no less than twenty different languages and were perfect specimens of their cultures: Negro wrestlers, Indian divers, Tartar bowmen, North African Moors, and Turks. They often accompanied the cardinal on his hunting expeditions. When he died in 1535, his "troop" carried his corpse the 200 miles from Itri to Rome, their diverse languages and violent gesticulations adding a certain dissonance to the general mourning.

Some of the cardinal's highly placed contemporaries added rhinos, camels, and giraffes of the Old World to their menageries, as well as jaguars, eagles, and llamas from the New World. Cortés discovered that the Aztec kings had assembled wild animals for display. The collection of the Grand Duke of Tuscany boasted an agouti and a type of West Indian weasel. Since many of the New World plants and animals appeared downright bizarre, collectors gathered them for their shock value. Monstrosities of all sorts played a starring role in the cabinets of curiosities that fashionable people assembled. A retrospective examination of thirteen menageries found "tiny jewel humming birds and gaudy grotesque toucans," gigantic teeth, and lots of rattlesnake rattles. Shells became so common an imported item by the eighteenth century that people decorated walls and clock stands with them.[8]

All these wild objects from the Western Hemisphere intruded on a European intellectual scene at first totally unprepared for them. Educated men and women in the sixteenth century were more enraptured with what ancient authors said about the flora and fauna of the earth than the actual specimens around them. They were immured in the texts of Pliny, Galen, and Dioscorides, natural philosophers of the first and second centuries. Their method of adding knowledge to the world's store was by extending the list of resemblances and associations that linked human beings to their natural environment.

Having contributed to the awakening of interest in natural phenomena, classical experts took a beating when their admirers found flaws in their texts. The sixteenth-century Flemish anatomist, Andreas Vesalius, for instance, discovered in his dissections of the human body that Galen had made flawed inferences about the human heart. He also saw that blood vessels originated in the heart, not the liver as Galen had said. Some traditionalists resented Vesalius's slur on Galen's reputation, but they couldn't stop the fact gathering. In the next generation, anatomy took another quantum leap forward when William Harvey set forth the principles for understanding how the heart pumps blood through the body.

Botanists confined themselves to the plants in their region and only slowly departed from the categories of Pliny and Dioscorides. The ancients described plants so generally that their sixteenth-century followers initially had little trouble placing American plants into their schema, but the very fact that the ancient authors were ignorant of the new specimens undermined the usefulness of their texts and, over time, their authority. Plants, of course, had medicinal uses that made them particularly interesting to human beings, with diverse peoples using bark or leaves of shrubs and trees to treat ailments. The ancients had compiled works about herbals containing the world's store of knowledge on plants with medicinal properties.

The effort to fit the natural plant life of the Americas into the categories passed down from Dioscorides encouraged some bota-nists to examine the new specimens in a very precise way. Without intending to disrupt comfortable assumptions, they described the novel objects from the New World with enough detail to frustrate incorporation of them into existing classifications. These careful comparisons set in motion an avalanche of botanical inquiries as people began to bestow upon the familiar, local vegetation the same exquisite attention that plants from the New World were receiving. A different kind of investigator emerged—one who preferred tramping in the woods to delving into the meaning of the ancient writers. People coined a new word for them: natural-ist. Without learning Greek, Latin, or Hebrew, these naturalists could pursue their intellectual impulses.

Those gazing upon a pineapple or palm tree began to insist that only a great artist like Leonardo da Vinci could do justice to the specificity of the spectacular beauty of these specimens. Only the artists' pens could capture the particulars that distin-guished one quadruped from another, one flowering shrub from its lookalike.[9] Jacopo Ligozzi, an Italian painter who lived at the turn of the seventeenth century, rendered animals with wonder-fully careful attention to anatomical detail. Ligozzi spent time painting at both the Hapsburg court in Vienna and the Medici court in Florence. His precision reflected a zeal that surpassed the mere goal of drawing trustworthy pictures of plant subjects. Ligozzi's precise sketches moved beyond prettiness and included the roots of plants. With detailed accounts and illustrations, the exotica of the New World eventually acquired separate identities.[10] In a similar vein, the Flemish artist Jan van Kessel's *Study of Butterfly and Insects*, for instance, rendered a lifelike wasp, mosquito, and beetle walking about a stem of grapes. Serious botanists benefitted from their work.

Slowly nature took on the role of instructor. Studying a particular shrub or flower became an end in itself. Patrons of the arts shifted some of their philanthropy to professorships in botany or zoology, or to the founding of museums and zoos. Enthusiasts founded academies devoted to the study of nature. The earliest still-surviving one dates from 1652 and became the Leopoldino, the national academy of Germany. It also pioneered publishing with the first scientific journal, *Ephemeriden*, twelve years later. The Royal Society of England, the Academy of Sciences of Paris, and the Royal Society of Sciences in Uppsala followed within a few years. Taking possession of the new phenomena intellectually was the work of several generations. Printed works did their part; so did collections. Museums had left their origins in the Greek muses of art and music to serve the public with access to the expanding knowledge about the natural world, now under persistent observation.

Botany was one of the first scientific inquiries to benefit from the Spanish explorations. Leonhart Fuchs, a German physician, established the first medicinal garden as part of his charge to reform the University of Tübingen in the 1530s. For the last thirty years of his life he worked to integrate the Greek medical tradition from Galen and Hippocrates with the new emphasis on cultivation and observation. During this same time Otto Brunsfels, working in Bern and Basel, published an herbarium with refined woodcuts based on keen examinations. Hieronymus Bock, another German physician of the early sixteenth century, started categorizing flowers by their similarities.

These botanists achieved a certain kind of disguised immortality through plants named after them. Thus Fuchs gave his name to the fuchsia found on Santo Domingo and Brunsfels to brunfelsia, a shrub sometimes called "yesterday, today, and tomorrow" to mark its three-day blooming period. Bock is remembered, if at all, because he named the vine pressed into service to produce the Rhine wine Riesling. Conrad Gessner, who climbed the mountains of his native Switzerland every summer in search of botanical specimens, gave his name to the genus *Gesneria*, a flowering sage. In his botanical writings he celebrated the beauty of nature and the physical benefits of outdoor excursions. Gessner, an expert in linguistics as well, wrote the first comprehensive text in zoology, covering quadrupeds, birds, and fish. Embellished with splendid illustrations, the first four volumes appeared from 1553 to 1558, the fifth on snakes following in 1587 after his death.

While some of the sixteenth- and seventeenth-century naturalists had clerical positions, many got their sinecures as royal doctors. Francisco Hernández de Toledo became the personal physician to Philip II, who sent him to Mexico in 1571 to do a comprehensive survey of plants of medicinal value in Spanish America. In seven years of work Toledo produced thousands of specimens and drawings of species even though his expedition had visited only Hispaniola, Cuba, and Central Mexico. Working with indigenous guides, Hernández located Indian herbal remedies and treated patients with them. To his astonishment his unlettered guides could accurately identify hundreds of plants. They were also capable of misleading him and frequently lied about the plants that they showed him.

Plagued by ill health as well as perfidious assistants, Hernández returned to Spain "laden with manuscripts, pots, barrels, sacks of seeds, and roots and casks with growing plants."[11] The latter he gave to the Alcázar gardens in Seville. He brought back extensive drawings and notes from the material that Franciscan friars had copied from Nahua scribes.[12] Although only some of Hernández's findings were published after his death in 1587, extracts from his writings in manuscript made their way into the burgeoning literature on botany in Europe.

Ulisse Aldrovandi, a contemporary of Hernández, dreamed of organizing expeditions to the New World, but fate intervened. At age twenty-seven he was labeled a heretic and placed under house arrest in Rome for a year. Here he cultivated the company of anyone who shared his zeal for botany, zoology, and geology. After abjuring his earlier heretical rejection of the doctrine of the Trinity, the church freed him to travel, and he went back to his native Bologna, where he got degrees in medicine and philosophy. While he taught philosophy, he also became the university's first professor of natural sciences. During the next fifty-five years, Aldrovandi pursued his passion for botany through expeditions to the mountains, coasts, countryside, and adjacent islands of Italy, earning for all this focused effort the title of "the Bolognese Aristotle."

A diplomatic stint in Spain introduced the Venetian nobleman Andrea Navagero to the refinements of Moorish gardens in southern Spain. He too received fresh reports on the exotic specimens that Spaniards were bringing back from the New World. Returning home with a cache of stories about the Spanish explorations, he designed the first private botanical garden in 1522.[13] This model no doubt influenced Aldrovandi, who laid out Bologna's public botanical garden, which still flourishes today. Aldrovandi assembled thousands of specimens of birds and other members of the animal kingdom for it. His drawings of eleven thousand items—animals, fruits, and minerals in addition to seven thousand dried and pasted plants—found their way into more than a dozen volumes. This record banished the inaccessibility of much of nature and brought it into the cozy confines of a museum. In keeping with that program, he arranged holdings according to a pleasing appearance. His collection included eight thousand illustrations, some by Ligozzi, as well as Aztec artifacts.[14]

Despite his scientific cast of mind, Aldrovandi lived in an age of portents, and he didn't mind using them to his advantage. An avid supplicant for gifts to finance his extravagant collecting, he exploited the sensation of a "dragon" (probably some kind of large lizard) found in the Bolognese countryside on the day in 1572 that Aldrovandi's cousin was invested as Pope Gregory XIII. Aldrovandi immediately put the creature on view in his museum, directed his full-time artist to render a likeness, and set about writing a history of the dragon. As word spread, speculation began as to what the discovery portended for the pope and his church. Hoping for papal support, Aldrovandi adroitly deflected attention from possible demonic messages to reasons why serpents—normally feetless—might acquire them.[15]

Not content merely to collect, Aldrovandi also wrote hundreds of essays and books on birds, insects, serpents, fish, and quadrupeds. His studies, most of them published after his death, were as eclectic as his collecting. He mixed information from personal observation, ancient texts, medieval tracts, fables, and emblems to probe the nature of things in this world as well as the imaginative uses to which diverse humans had put them. Aldrovandi introduced the classification of mammals as solid-hoofed, cloven-hoofed, and clawed. Despite their absence of hooves, he correctly joined whales to the mammal group.[16]

Before his death in 1605, Aldrovandi arranged to give his vast collection to the Senate of Bologna, which moved it to a palazzo in the center of the city. Without undue modesty, Aldrovandi called his museum the eighth wonder of the world. A stipulation of his bequest was that the Senate publish his voluminous manuscripts, and it did, until their out-of-dateness became conspicuous in the 1670s.

Bologna was not the only city with a famous collection; Naples, Rome, Verona, and Milan had their own "wonder rooms," "cabinets of curiosity," or "theaters of nature," chockablock full. Shells, stuffed animals, feather headdresses, embalmed fish, flags, and framed butterflies hung from walls or were thrust into view on elaborate stands. Even more important, the catalogues of these collections became repositories of natural history, detailing how the novel objects had been acquired and locating them within pertinent literature.[17]

Gardeners and their more intellectual brethren, horticulturalists, cultivated specimens from the New World that yielded hundreds of flowering shrubs and trees for European parks. Another pioneering naturalist, Carolus Clusius, studied medicine but practiced little, devoting his time instead to botany. He moved from the imperial court in Vienna to Frankfort to Leiden, where he created a famous botanical garden in the closing decade of the sixteenth century. A close observer and careful note taker, he turned out richly illustrated books. Like many of his peers, Clusius was a true polymath, reading in eight languages while pursuing the law, philosophy, history, cartography, numismatics, epigraphy, zoology, and mineralogy.[18]

When the Holy Roman Emperor's ambassador to the Ottoman sultan sent Clusius some tulip bulbs, he began cultivating them, only to set off a rage. The Dutch went mad for these handsome flowers, many with splendid feathery petals and vivid stripes. A heady combination of rarity and beauty produced a mania that climaxed in Europe's first commercial bubble. For two years in the 1630s, people bid up the price of tulip bulbs. As is the way with bubbles, this one burst, leaving lots of people poorer, maybe even wiser, and with colorful annual tulip blooms to remind them of their folly. Meanwhile, Clusius had laid the basis for the tulip trade, which the Dutch dominate to this day. Clusius also tramped through the Alps in search of plants, many of which now bear his name. He undertook the first study of mushrooms from Central Europe. He, with several of his peers, was responsible for starting in the second half of the sixteenth century cooperative printing projects that pooled data about the living world.

European botanists were aware of the reports on American plants to be found in the works of Oviedo and José de Acosta, but it was not until the seventeenth century, and often spurred by new botanical travels, that New World flora was incorporated into the authoritative texts emanating from Dutch presses. At the same time, many American specimens could be found in the gardens of European amateurs and specialists alike.[19] Botanists called their compendiums florals. With a separate category for their published texts, they slowly detached themselves from the concerns of physicians and apothecaries.

At the same time that botanists were pondering whether New World specimens were but variations of known plants, Europeans got a chance to taste some New World edibles—and the denizens of the Americas, some Old World ones. On his second trip, Columbus had brought seeds for all the Spanish fruits and vegetables that he hadn't seen in the Caribbean on his first visit. His successors continued this work of mingling plants that grew in the once-separated worlds. Spaniards and Portuguese brought bananas, lemons, oranges, pomegranates, figs, dates, and coconuts, the latter found in the Philippines. From the New World, Europeans received a variety of squashes, cocoa, and tobacco, and that treasured delicacy, dried pineapple.

Gonzalo Fernández de Oviedo had included in his history pictures of the corn he ate in America. It became so common in European vegetable gardens that when Fuchs wrote about it a century later, he placed its origins in Turkey. Europeans proved more resistant to the potato. Clusius did note its fecundity, even though he concentrated on his own backyard, so to speak. It took more than another century for this nourishing tuber, already present in some continental gardens, to reach the poor of the British Isles. The harvest bounty of the humble potato finally won over the Irish long after Walter Raleigh had introduced it there on his estates.

The range of European vegetables and fruits was far broader than that in the Western Hemisphere, but a few American staples like potatoes, beans, and corn were destined to play a major role in feeding Europeans after the initial resistance to their novelty eroded. These hardy foods had different cultivation requirements, which meant they could enhance Europe's collective food supply. All could be grown in places inhospitable to the grains Europeans depended upon. Corn, for example, could grow where it was too wet for wheat and too dry for rice, and it yielded twice as much food per acre. These carbohydrates added critical sources of nutrition to a continent still visited by famines.

The potato was even richer than corn in calories and could thrive in very small plots. Even more remarkable, potatoes yielded two to three times more bushels per acre than wheat or barley. They could be stored through the winter and didn't demand much in the way of cultivation. Potatoes thrived at high altitudes, helping the Spanish feed the silver miners of Potosí. The same virtue enabled Chinese peasants to move into hill country to flee government tax collectors.[20] When invading armies burned crops to the ground, potatoes remained hidden in the earth.

People are amazingly slow to adopt strange foods, however many benefits they bring. Those in the know could marvel at the New World's Jerusalem artichokes, strawberries, agaves, and prickly pears, but most found them just peculiar. In the seventeenth century there was a sharp increase in plantings of such novelties as sunflowers, nasturtiums, morning glories, passionflowers, dahlias, and petunias because horticultural enthusiasts now outnumbered the physicians and apothecaries who had dominated gardening before.[21]

With more of an affinity for animals, Count John Maurice of the Netherlands made the first systematic study of New World wildlife in 1637. Maurice was strategically placed to make a singular contribution to natural philosophy. Born into a privileged family, he was the Dutch West India Company's choice to govern their holdings in Brazil after they seized them from the Portuguese in 1630. In that twenty-four-year period, Count Maurice laid out a splendid town with public buildings and gardens, which he modestly named Mauritsstad. While exercising his talents as an administrator, he also demonstrated his military prowess with further conquests in Brazil and on the west coast of Africa. He initiated public housing projects and extended tolerance to the mixture of Jews, Catholics, and Protestants that made up the European contingent of his province. Eventually, the Dutch West India Company threatened to curtail his generous spending, and he returned home.[22]

From his own funds Maurice financed what became the first thorough descriptive and visual record of the natural history in any part of America. Breaking from tradition, he gave equal attention to plants without medicinal properties. Among the forty-six scholars and craftsmen he had taken with him to Brazil were six artists. Two of them, Albert Eckhout and Frans Post, made careers of painting scenes and objects from the Americas long after they had returned home. Post specialized in sweeping landscapes where tropical greenery and bizarre animals dwarfed figures of native men and women and enslaved Africans. Eckhout depicted the Tapuya Indians of Brazil with an accuracy that shocked viewers back home, neither idealizing the Indians nor pandering to morbid curiosity. Luscious watermelons, cashews, passionflowers, and pineapples became motifs, gleaned from paintings and specimens, in a series of life-size Gobelins tapestries that Maurice presented to Louis XIV in 1678.[23]

Count Maurice's scientific forays gave a boost to the study of birds as well. Blue jays, woodpeckers, hawks, crested curassows, and eagles had made the return voyage across the Atlantic with the earliest explorers. Until 1520, those who knew anything about the New World still thought it a part of Asia, so the pioneer ornithologist Pierre Belon gave both the duck and the turkey Asian origins. He reproduced drawings of the "Muscovy" duck and described the turkey as from India, or "d'Inde," from which came *dinde*, the French word for turkey. Sailors returning from Brazil enlarged Belon's collection with carcasses of birds from which he drew pictures of the macaw and toucan. But soon live birds began surviving the trip east to fill European aviaries, while the duck and turkey added variety to the fowl course at European banquets.[24]

John White, whom we met when he participated in the ill-fated English settlement in Roanoke, had sketched animal life in both the West Indies and the North American continent, but the expeditions of Count Maurice marked the beginning of organized studies of its birds and mammals. A generation later, Mark Catesby brought Maurice's systematic approach to the wildlife of North America. "My curiosity was such," Catesby commented in recognition of its infectious nature, "that not being content with contemplating the product of our own country, I soon imbibed a passionate desire of viewing as well the animal and vegetable production in their native countries." This passion took him to Virginia and the Carolinas. Seeing his collection, his friends raised the money to send him back across the Atlantic to collect further specimens. Catesby was both determined and talented. He taught himself to make engravings and supervised the hand coloring for his landmark study, *The Natural History of Carolina, Florida and the Bahama Islands.*[25]

Leiden native Jan Swammerdam began as an amateur collector of coins, fossils, and insects. After receiving his medical degree in 1667, he traveled through Europe in pursuit of specimens to slake his appetite for information about bugs. In the next seven years Swammerdam published three volumes on his research, challenging Aristotle's claim that insects were too beneath one's dignity to merit the study one might bestow on fish or snakes. Swammerdam felt differently and conducted experiments to show the similarities between the development of insects and other animals. He proved that insects like caterpillars and butterflies didn't undergo a change of type but rather proceeded through different life stages from larvae to mature insects. The reproductive processes of the honeybee eluded him, but he managed to locate them in wasps, ants, dragonflies, snails, worms, and butterflies.

What was even more important to investigations of nature's phenomena was Swammerdam's use of the microscope in dissections and his experiments with frogs. He demonstrated how the brain worked through the nerves to move muscles. Swammerdam also had a talent for creating ingenious techniques, as when he injected wax into blood vessels to make them visible. All of this was virgin territory when he began his work. Many of his methods remained standard well into the next century. Years later, neuroscientists would draw upon his experiments to make the precise connection between behavior and stimulus. Swammerdam himself saw divine power in nature. His discoveries once prompted him to exult, "Oh wondrous God! Who would not know you from this and, knowing you love you!"[26]

Swammerdam corresponded with another naturalist, Melchisédech Thévenot, who was also putting his boundless curiosity, unflagging energy, and generous fortune to the cause of science. A diplomatic career nurtured Thévenot's intellectual passions, taking him to Genoa and Rome and their famous libraries. His studies resulted in *Relations de divers voyages curieux*. Published between 1663 and 1672, the four volumes presented a compendium of travel accounts that he had gathered and translated.[27] He also performed experiments, demonstrating in a dissection of a human brain that atmospheric pulsations had something to do with human respiration.

As a patron of science, Thévenot helped found the French Academy of Sciences. His popular work on the art of swimming reached across the Atlantic in the next century to prompt a young Benjamin Franklin to become a lifelong swimmer. Thévenot recommended the use of lemon juice to counteract scurvy and introduced ipecac as an emetic for dysentery. His invention of the bubble level became a boon to carpenters, bricklayers, stonemasons, surveyors, and anyone else who wants to level a surface. Gottfried Wilhelm Leibniz, the German mathematician and theologian, called him a universal man, noting that "nothing escapes his curiosity."[28]

These amateurs were ingenious. Swammerdam and Thévenot were exemplars of a potent new form of inquiry: how nature works in bodies and plants. But they were not alone. The fourth and fifth decades of the seventeenth century produced a cluster of brilliant naturalists who took advantage of autopsies and microscopes to lay bare the function of internal organs. They worked in Italy and England in addition to France and the Netherlands. Their research laid the foundation for an understanding of reproduction in animals, insects, plants, and human beings. Cumulatively, their work broke decisively with traditional anatomy and its hallowed ancient texts. A different focus emerged along with a change of sensibilities.

Nehemiah Grew, like Swammerdam, took a degree in medicine at the University of Leiden. He returned home to London to practice as a physician, but found time a decade later in 1682 to publish his *Anatomy of Plants* covering roots, trunks, leaves, flowers, fruits, and seeds, all handsomely illustrated in eighty-two plates.

He was the first to identify pollen. Grew shared his work with Marcello Malpighi, a native of Bologna, an abundantly talented botanist and sketch artist. In his *Anatome Plantarum* Malpighi made the first detailed drawings of the organs of flowers. He also worked with animals, using autopsies as well as vivisections to reveal the secrets of the kidneys and liver, and he did ground-breaking work on skin pigment.

Malpighi's work on the silkworm brought him an international reputation. His and Grew's contemporary, Regnier de Graaf, another graduate of the Leiden medical school, made human sexual organs his focus of attention. At the time men and women were assumed to have analogous testicles and semen; eggs and spermatozoa were not identified until later in the century. De Graaf figured out the function of the fallopian tube, referring in his description to its "very elegant shape." More surprising was his hypothesis that the ovary actually released the egg, though no one observed this critical detail until the nineteenth century.[29]

A self-taught scientist, the Dutchman Antonie van Leeuwenhoek came to science by way of a fascination with magnification. He was reared for the linen trade, where the regular inspection of cloths involved the use of a simple device that could enlarge the view of the thread three times. This instrument captured his imagination. From the ages of twenty to forty he experimented with spinning thin threads of glass under heat and made dozens of devices framing tiny glass beads in copper casing. His microscopes finally achieved magnification to the power of a thousand. Leeuwenhoek never disclosed his ingenious technique, so no one succeeded in duplicating it in his lifetime, but he let others see his world of invisible phenomena if they shared in his search for the truth about God's creations.

With increasingly powerful microscopes, Leeuwenhoek examined the hairs from a sheep, the legs of lice, the stingers of bees, plant seeds, and even the eye of an ox. Though not well educated, he had the native gifts of a sleuth, spending hours upon hours gazing at the objects that slid under his lenses, all the time making careful notes of what he saw. One day Leeuwenhoek put a drop of rainwater under his microscope and was astounded to see thousands of "beasts" darting about in the water. He had discovered microbes! He brought home water from a local pond and reported seeing "floating therein divers earthy particles" moving incredibly fast. He was actually observing bacteria without knowing what they were. He examined the tail of a fish and saw the capillary blood vessels through which blood goes from the arteries to the veins.

De Graaf, who was a corresponding member of the English Royal Society, recognized Leeuwenhoek's genius and encouraged him to write the society about his findings. He would do so for the remaining fifty years of his life in long, rambling letters reporting on the increasingly complex investigations he was making of that new world previously inaccessible to unaided human eyes. When he wrote the Royal Society about the little creatures dashing about in water, the members tried to replicate his examination and failed. Grew, who had become the secretary of the society, dispatched a delegation to Delft to look through one of Leeuwenhoek's microscopes for confirmation. The visitors were convinced and envious, but Leeuwenhoek remained unyielding in the matter of selling them one of his precious microscopes.

Both indefatigable and eccentric, Leeuwenhoek drew luminaries like Tsar Peter the Great to his native Delft to look at the muscle fibers, bees, spermatozoa, blood vessels, and bacteria that he placed for them under his magnificent lenses. He was the person who first realized that the fertilization of the egg took place through its union with sperm. With enough dedication to take the place of an education, Leeuwenhoek brought coherence to microscope studies.[30]

The Royal Society of London became critical to the work of

naturalists in the seventeenth century. It reached out to those doing research across the continent. As in the case of Malpighi's studies, it published work and offered judges to confirm experimental results, as in the case of Leeuwenhoek. When Swammerdam claimed priority for his work on the ovary after the publication of de Graaf's book, the two naturalists put the dispute before the society, which alas did not resolved it before de Graaf's death at age thirty-two in 1673.

These scattered experimental initiatives found a theorist and propagandist in Francis Bacon. An English courtier and judge, Bacon developed a passion for studying nature through observation and experimentation. We might ask how else would one study nature. In championing a new approach—one in which natural objects were looked at with questioning eyes rather than explained by Aristotelian deduction—Bacon had many goals. One of them was the advancement of learning, the title of his famous 1605 treatise. He also thought a careful investigation of natural objects could reform manners! Having heard a lot of bombast in his long career at court, Bacon came to detest opinion—what today we might call ideology. For him, opinion only promoted heated conversations, never truth-seeking.

Examining natural phenomena, reasoned Bacon, would train the senses and feed the mind because nature was a master disciplinarian. If you had opinions about nature, they could be tested, unlike notions about why people competed with one another or what might explain the hostility of a foe. When you proposed some erroneous hypothesis, nature would fight back. If you opined that frozen objects could be melted by the sound of lively music, for instance, you could learn forthwith that this was not true. Bacon in fact died of pneumonia caught while experimenting with ice. He looked to experiments to arbitrate disputes and thus avoid lengthy wrangling about aspects of nature, while opinions about politics or morals circled unceasingly because there was rarely a way to disprove them.

Aristotle, who had dominated discussions about nature for twenty centuries, had viewed the world through the relationship of matter to form. The reactions of matter differed, he taught, according to its essence. The four basic elements of air, water, earth, and fire conveyed the qualities of dry, wet, cold, and hot. Heavy objects fell to the ground, according to Aristotle, because it was an inherent quality of their heaviness. When his contemporaries repeated such propositions, Bacon tapped his considerable reservoir of invective to excoriate them. Medieval science, following Aristotle, concentrated on what was normal, regular, and universal about the created universe, seeing the abnormal and aberrant as accidents of little value to their inquiries. Even Galileo stayed within this Aristotelian framework, though he replaced quantitative for qualitative reasoning.[31]

Directly challenging Aristotle's focus on regularity, Bacon wanted investigators to take on the challenge of explaining errant items such as children born with six fingers, Siamese twins, dogs with extra tails, optical illusions. His contemporary Leibniz concurred. He favored a public exhibition of "all sort of optical wonders ... unusual and rare animals ... extraordinary rope-dancers ... artificial meteors ... ballets of horses, museums of rarities" along with "calculating machines, the air pump, an anatomical theater, and the oscillations of the pendulum."[32] Bacon's idea was acquiring legs, at least in the realm of ideas.

Usually scrutinized as portents, such marvels, if carefully studied, Bacon thought, should reveal how these deviations conformed to nature's rules of cause and effect. Laying bare the causes of freaks of nature, aberrant phenomena, and rare sightings would impress doubters. Though they would be difficult to pull off, such proofs would greatly enhance the reputation of empirical studies, an enduring goal of Bacon's.

Bacon disliked quick assessments. He viewed them like opinions—too easily acquired and too weakly based to be convincing. He believed that taking on difficult assignments would stay the

hand of experts prone to generalize prematurely. Only explanations heavy with factual observations and battered with dili-
gent questioning should be allowed into the knowledge archive.
Deviations in nature occurred. Thus they should be studied, not
shunted to the side as boring exceptions that threatened rules.
Natural history should inform natural philosophy, not the reverse,
as in the Aristotelian Scholastic tradition. Out of this came a new
meaning for the venerable legal term "fact"—particulars gleaned
from experience and shorn of theory.[33]

Bacon challenged reigning conventions in several critical ways.
He introduced a systematic skepticism about how people formed
opinions, sarcastically summarizing them in his study, the *New
Organum*. There were the "Idols of the Tribe," ideas transmitted
to children by their society; "Idols of the Cave," notions gained
through personal experience; "Idols of the Market Place," opinions that came from social intercourse; and "Idols of the Theater,"
for Bacon the received wisdom of the day, which he described as
so many stage plays perpetuated by "tradition, credulity, and neg-
ligence."[34] These kinds of tropes almost made Bacon himself into
an idol for the like-minded; he certainly was the powerful patron
of empiricism for those hacking their way through tradition to
follow the path of open investigations.

Going against the grain by extolling useful knowledge, Bacon
sought practical outcomes rather than moral tutoring from studying nature. This goal brought natural philosophy down from
its lofty perch and positioned it in the midst of active men and
women. Bacon also subscribed to the wide dissemination of
experimental results. These emphases on usefulness, communication, and transparency tilted against the hierarchical structure of
his society and its institutions. Knowledge had long been treated
as a body of secrets to be passed on to select groups only. Though
Bacon was no democrat, his faith in spreading information about
research represented a pathbreaking departure. Telling others

about one's studies of nature widened the ambit of participation, making it possible for strangers to share their insights.

For Bacon and his many followers in succeeding generations, getting out the word of new experiments would attract inquirers from outside the upper classes. Having to know Latin had acted as a high entry fee when natural philosophy depended upon ancient texts. Findings written up in vernacular languages attracted those men and women with an aptitude for natural philosophy. As empiricism gained a greater and greater hold on the European imagination, Baconians assumed that metaphysical speculation about unknowable and imponderable subjects would begin to wither in the face of the painstaking accumulation of facts that had survived rigorous tests.[35] In this, they were a tad optimistic, but over time it did happen.

Bacon collected targets the way a cook assembles recipes. Because he had a fully articulated and comprehensive theory of knowledge, his ideas butted up against an array of preconceived notions. A powerful one was that of degeneration. The centrality of ancient Greek philosophy and the biblical origins of Christianity encouraged people to look backward to these golden ages with the implication that all subsequent societies were but pale imitations. This sense of degeneration drew strength from the human life cycle of growth to maturity to the waning powers of old age. A variant on this conviction was that of cyclical patterns in history as societies grew and flourished and declined, which in turn called forth a rebirth of vigor, usually in another area. Bacon found these deeply entrenched ideas unacceptable. They induced an exaggerated estimate of ancient glories and a refusal to see the contemporary progress in the growth of knowledge about nature that was so obvious to him.[36]

Bacon recognized the crucial role that the Spanish discoveries had played in stimulating interest in natural objects. He credited the ventures with arousing people from the torpor of tradition.

"Distant voyages and travels," he wrote, "have brought to light many things in nature, which may throw fresh light on human philosophy and science and correct by experience the opinions and conjecture of the ancients."[37] The capture of animals and plants with pen or cage throughout the sixteenth and seventeenth centuries had created a vast archive from which scholars could put together more and more authoritative natural histories. The objects retrieved from the New World also suggested to Bacon a more specific weapon in his fight against the intellectual tyranny of the ancients.

According to Bacon there were four supports for philosophical learning: a library, botanical and zoological gardens, a cabinet of curiosities, and a collection of instruments, furnaces, and vessels for experimentation. His ideas got a typical English twist in the eighteenth century when enterprising lecturers went from town to town loaded with the equipment essential to demonstrating to eager lay audiences the latest scientific and technological advances, especially Newtonian mechanics. Charging handsome fees, these travelers with their showcases exemplified the great Baconian goal of stimulating knowledge by sharing it.[38]

The eminent English natural philosopher Robert Boyle was a splendid exemplar of the new investigative spirit. At age fourteen he traveled with a tutor to Galileo's home in Florence. A decade earlier, the Italian Inquisition had forced the stargazing Galileo to abjure his belief that the sun is the central body around which the earth and the other planets revolve. Galileo had already distinguished himself as a mathematician and astronomer when the church put him under house arrest for his heliocentric views. The man who conceived the accurate laws of motion left a profound impression on the young Boyle. In a kind of passing of batons, Boyle was born in 1627, nine months after Bacon's death, and Newton was born in 1642, the year that Galileo died.

Boyle was a member of the "Invisible College," a band of like-minded gentlemen who met in London and Oxford to discuss mechanics, or how things moved. As the independently wealthy son of an earl, Boyle could devote his life to physics and chemistry, and he did. Aristotle, that ancient authority in all matters philosophical, had said that nature abhorred a vacuum. Willing—perhaps eager—to prove Aristotle wrong, Galileo had experimented with suction pumps. Having hitched his wagon to Galileo's star, Boyle used an air pump and a bell jar to prove conclusively the existence of a vacuum. The stunning conclusion: atmosphere had weight. With this knowledge, it was possible for gifted mechanics to design the first steam engines a half century later.

Best known for his work on expanding gases, Boyle took on just about everything, from sound, electricity, combustion, respiration, color, and hydrostatics to the establishment of chemistry through a rigorous study of the composition of substances. An indefatigable researcher, he had the benefit of servants at his disposal who could help him. Many a time they assisted in late night observations of phosphors gleaned from putrefied fish or urine. These experiments were so successful that one of Boyle's assistants set up a factory to produce phosphorus, whose luminescence gave it a commercial value. Like Bacon, Boyle distrusted facile hypotheses, a point of view he stressed in his *Sceptical Chymist*, published in 1661.

Boyle even found time to write a book on natural history for the use of travelers and navigators. These how-to manuals helped peripatetic Europeans organize what they saw on their trips. Advocating order, Boyle set out categories such as stature, shape, skin color, food, habits, occupation, art, virtues, vices, and wit to guide the attentive traveler. With subscription libraries and book clubs flourishing, the community of natural philosophers succeeded in recruiting a whole cadre of amateurs.[39] Deeply religious, Boyle argued in other writings that the study of God's creation was an act of devotion. The conviction that God's glory was revealed in the smallest details of nature became a new frontier in the spiritual realm, as empirical studies won new converts with experiments showing how nature actually operated in the world.

Boyle employed Robert Hooke, one of the polymaths of the seventeenth century. Hooke anticipated some of Newton's ideas about gravity, improved on the mechanics of timekeeping, demonstrated how dogs breathed, laid out plans for London after its great fire of 1666, and patented a number of inventions in elasticity, optics, and barometry. He designed the air pump that Boyle used in his famous experiments. For the last forty years of his life Hooke served as the curator of experiments for the Royal Society, a post that placed him at the center of scientific work in England as well as the disputes that the new work generated.[40]

Collecting and investigating continued apace, the collecting allowing those without the expertise of experimenters to participate in the burgeoning fields of botany, zoology, mechanics, and physics. The popularity of collecting led to a new status: that of the connoisseur. Often the agents of the wealthy, connoisseurs acquired both taste and knowledge through personal experience. They turned their "eye" for quality into an asset. Two generations of Tradescants in England established one of the great agglomerations of wondrous objects in all of Europe in the first half of the seventeenth century. Using connections with a handful of great noblemen, the senior John Tradescant acted as the indefatigable scout and purchasing agent for several aristocratic patrons. He designed both gardens and cabinets of curiosities for his clients and in the process got to keep duplicates given to him.

Tradescant aimed at a comprehensiveness of both objects and global regions, and he passed on this obsession to his son. When John Jr. sought to prepare a catalogue for the family museum, he got caught in the toils of a wily lawyer, Elias Ashmole. Ashmole wanted to perpetuate his own fame in a gift to Oxford, so he took possession of the collection after his client's death. The Ashmolean Museum is a living testament to the Tradescants' avidity for the rare and beautiful, whether a bird of paradise shrub or the wondrous dodo bird that could not fly.[41]

A half century later another great collector, Hans Sloan, acquired a great trove of manuscripts, seals, coins, medals, cameos, stuffed animals, and drawings. Sloan, a physician, made a fortune when he brought back Peruvian bark, the source of quinine, from a Caribbean visit and then prescribed it for various ailments. With his riches he amassed, contemporaries said, more priceless objects than any other private individual. He also bought collections to complement his own. His passion did not always produce satisfaction, as Sloan discovered when his famous visitor, George Frederic Handel, put a buttered muffin on a rare manuscript. After Sloan's death, Parliament bought his treasures to lay the foundation for the British Museum; the collecting of Francesco I, Grand Duke of Tuscany, gave the Uffizi Gallery in Florence its start.[42]

Collections, begun with a strong push from vanity, became part of a Europe-wide turn to empirical science. The creation of knowledge went from exploration to travel to collecting to permanent accessibility in public museums that doubled as scientific resource and general entertainment. By the seventeenth century, catalogues of collections acted as an additional resource for studying natural phenomena. By the eighteenth, most royal menageries became public zoological gardens, or zoos.

Slowly knowledge supplanted utility. As expert attention turned from the use of plants to the plants themselves, questions arose about what constituted a family of plants, how the plants replicated themselves, what role the parts of flowers played in that process, and what stages formed the life cycle of a plant. A reciprocal interaction began. The emphasis on observation gave an impetus to finding new specimens, and these in turn provoked more attentive examinations. A passion for collecting could now be joined to the excitement of travel, if only to remote spots in one's own country. The new activities of searching, studying, categorizing, and displaying natural objects found many champions in people of a humanist cast of mind.

The specimens of the Western Hemisphere caused cognitive dissonance for those tethered to ancient botanical texts. Europeans had been reared in a tradition that rooted the origins of all life in the Genesis account of the Creation and Noah's Ark. This of course was even more of a cerebral headache when it came to fitting in the human communities that Columbus encountered. But what one generation experiences as a shock, the next deals with as a problem. Slowly, during the sixteenth and into the seventeenth century, medicine lost its grip on botany. Collectors let physicians cultivate their medicinal plants and theologians tussle with issues about origins while they pursued their scrutiny of natural objects, establishing new standards for authoritative texts.

Both formal and informal associations allowed merchants, professional men, and leisured persons to share their excitement about nature in sociable gatherings that eroded the barriers separating noblemen from commoners in Europe's aristocratic societies. Naturalists often got together to read each other's writings and confirm each other's experimental data. Coffee shops and private libraries became sites for wide-ranging intellectual discussions. Books spurred conversation; they also connected people who might never meet each other or might not even live at the same time. Slowly a new authority was being created—that of a community of natural philosophers who could form a consensus of experts. Keepers of good order decried the indiscriminate mixing of gentleman with members of lower classes. Affinities based on common interests appeared on the social horizon as a real threat to the established system of statuses.[43]

England was peculiarly hospitable to the new mode of inquiry. Boyle's "Invisible College" became the nucleus for the English

Royal Society. Rigor became the watchword for its research, something van Leeuwenhoek discovered when he sent his findings about single-cell organisms to the Royal Society. As with other European academic societies, its members were elected for their contributions to the burgeoning portfolio of scientific inquiries, but political connections and social prominence helped a candidate get in.

Boyle also set a standard for the publication of experimental procedures. They had to be clear and complete enough to easily be replicated.[44] These academies began to draw in the intellectually curious, displacing over time the authority of men from the universities, where Aristotelian Scholasticism often reigned unchallenged. In the Baconian spirit of producing useful knowledge and, probably, to justify its royal support, the Royal Society initially surveyed farming practices across England. It sponsored, among other things, a study of the use of the potato as food. Far more important, it brought together in the same room the people who were most engaged in solving physical, mechanical, and mathematical problems.[45]

There was a downside to this openness. The English Royal Society regularly published philosophical transactions, which had the unintended consequences of furnishing quacks and charlatans with an archive of experimental data that could be used to sell elixirs of various sorts. Newspaper ads featured cures based on ginned-up versions of society experiments.

The potato needed more than Royal Society support to gain popularity. Rumors arose that its tubers spread leprosy or, more pleasantly, were an aphrodisiac. Its amazing yield recommended potatoes to King Frederick the Great of Prussia, who ordered his people to plant them during the famine of 1744, but it was Antoine-Augustin Parmentier who finally delivered the tuber from suspicion. Forced to eat nothing but potatoes while a Prussian prisoner during the Seven Years' War, he became an advocate of the despised vegetable. When he returned to his vocation as a pharmacist in Paris, Parmentier mounted one of the world's most successful advertising campaigns. He put security guards around his potato plot, then removed them when he thought he had sufficiently intimated their value to potential (and hungry) thieves. He entertained guests with all-potato menus. Thomas Jefferson, then the United States foreign minister to France, supposedly brought back the recipe for French fries, which he served over the next decade at the White House.[46]

The stars and planets had long appealed to men and women, as illustrated by the myth of Icarus making wings to carry him to the heavens. An old belief held that things below the moon—sublunar—operated differently from those on the planets. But they were so far away, it was hard to tell. Johannes Kepler made a critical move in astronomy in the early seventeenth century. Blocked by his inability to perceive planets directly, he devised an indirect approach. Heavenly bodies could be studied with mathematical precision, he posited, by examining their shadows and the refraction of light. Optics would do the work of space travel.[47] At the end of the century Isaac Newton startled the philosophical community when he demonstrated that celestial bodies, as well as the earth, were subject to gravitational pull. All obeyed universal laws.

The laws governing stars and planets could be expressed mathematically. When Newton published his *Principia* in 1687, many could not understand the mathematics involved in his proof of gravity's pull, but all educated people stood in awe of the human mind that could define laws of motion that applied on earth as well as in the heavens. The poet Alexander Pope captured the sentiment with his famous couplet intended for Newton's epitaph:

Nature and nature's Laws lay hid in Night;
God said, Let Newton be! And all was Light.

Despite English enthusiasm, French mathematicians were reluctant to embrace Newtonian physics or Bacon's insistence on empiricism. Bacon had argued that experiments, not conjectures, were the linchpins for knowledge of the natural world. His intellectual seeds fell on very fruitful ground in his own country. Due to the open character of English public life, knowledge moved from the esoteric investigations of natural philosophers to a broader community of the scientifically alert. Experience with air pressure, vacuums, and pumps became part of a broadly shared scientific culture that reached out to craftsmen and manufacturers in addition to those of leisure who cultivated knowledge.[48]

The free circulation of ideas through publications and discussions, along with the easy mixing of ordinary citizens with members of the educated elite, created a broad receptivity to the novel propositions that were challenging centuries of learning. Attitudes changed slowly, but each successive generation coming into adulthood shepherded the new ideas' passage from shocking to puzzling to acceptable.

Twentieth-century historians of science were chastened to discover that the Newton who discovered gravity also practiced alchemy, the effort to turn base metal into gold and silver. Magic—influencing actions from a distance—like making a person sick by sticking pins in an effigy or curing people with words continued to attract the attention of smart people, much as astrology does today.[49] They burned witches in Massachusetts and prosecuted them elsewhere well into the eighteenth century.[50] Because we know that experiments came to dominate the study of nature, we assume that they swept away the field of alternatives right away. But they didn't. Nor should we be surprised, since school boards in the United States are still debating whether or not to teach evolution 150 years after Charles Darwin published his *Origin of Species*.

The strand of curiosity that the New World explorations cul-

tivated represents an unpretentious, collateral line to the dazzling work in astronomy. The study of living forms—the plants
and animals with their intertwined human ecologies—attracted
its own researchers. Toucans, potatoes, pineapples, and armadillos may have struck people as strange, but studying them was
well within the capacity of many educated Europeans. American
flora and fauna encouraged men and women to take a second look
at the plants and animals that they had long taken for granted.
Lying two centuries ahead was another English scientist who
would show that even the innocent observation of flora and fauna
and their fossils could detonate an intellectual bombshell. Darwin would startle the world with his conclusions about the origin
of species even more than had the physicist Newton with his laws
of gravity.

The reversal of fortunes for the investigative spirit can be
captured in two opposite pieces of advice: "do not seek to know
high things" and "dare to know." The former originated with St.
Augustine and went through various restatements. In antiquity,
the myth of Icarus plummeting into the sea when the sun melted
the wax of his wings expressed a similar thought: one shouldn't
seek after knowledge beyond human ken. One should not desire
to know what is not directly given, a thought very alien to modern
people, but deeply felt centuries earlier.[51]

"Dare to know" came from Horace, a first-century Roman
poet. During the course of the seventeenth century it became a
kind of rallying cry among the hundreds of amateurs exploring
the myriad objects offered up by nature. Icarus—and even Prometheus, who was eternally punished for stealing fire from the
gods—became heroes for their risk taking. Like Icarus and Prometheus, Galileo and Columbus dared to move beyond the limits
set up to corral human imagination.[52] What had started out as
"the mysteries of nature" slipped almost invisibly into items on a
scientific agenda. The biggest intellectual somersault came when

examining objects replaced interpreting ancient texts. The idea of wonder in which one was immobilized by awe at what wasn't known—and presumably couldn't be known—gave way to persistent inquisitiveness.

Along the way, natural philosophers came to distinguish between appearance and reality. All was not what it appeared to be, whether one went from cause to effect, like the physicists, or effect to cause, as the medical men did. A new sophistication took hold that taught that things weren't always what the senses reported. Unlike believers in the occult who dwelt on the hocus-pocus of the hidden, natural philosophers were reaching beyond data gleaned from the senses for a reality beneath appearances that could only be grasped after intense questioning.[53] Invisible forces were difficult to find and measure, but that did not make them either mysteries or carriers of coded, providential messages. And the process of demonstrating this produced pleasure, at least according to the great philosopher Thomas Hobbes, who exclaimed over "the perseverance of delight in the continual and indefatigable generation of Knowledge, [which] exceedeth the short vehemence of any carnall Pleasure."[54]

Increasingly the world and all its living forms became an object to be interrogated and studied rather than a divinely given place for human habitation to be praised and revered. "Because that's the way God created it" no longer sufficed as an answer to a probing question about how things worked. The word "nature" took on new meaning suggesting, after Newton's great accomplishment in explaining the universal laws of gravitation, an orderly universe. Its secrets became challenging puzzles. A sense of wonder was now engendered by explanations of phenomena and not just the phenomena themselves.

Experts began to part from casual investigators through the serious application of their time to a single line of inquiry. As natural philosophers began to work with telescopes and microscopes, they came to appreciate the vast field of the not-yet-known. The interiors of Africa and South and North America remained terra incognita. Plant species still thrived outside the botanists' gardens. Animals, even human societies, remained secluded from the scientific eye. But there was plenty of time to change all this. One of the charges against inquisitiveness was that it was unbounded. The critics were correct. Once tolerated, then nurtured by discoveries and enhanced by new instruments, it knew no bounds.

CHAPTER FIVE

AMATEUR NATURALISTS
YIELD TO EXPERTS

Carl Linnaeus dressed as a Laplander, frontispiece of his
Flora Lapponica, *from a portrait by Martin Hoffman, 1737.*

In the twenty-first century, we know that if we don't know something about a natural phenomenon, some expert does. Three centuries ago that was not the case. Talented amateurs might be the first to describe an ocean tide or a rare flower. Discoveries provoked delight. Yet the knowledge gained produced ever more questions, pushing out the frontiers of knowledge. It was like hiking higher and higher up a mountain and seeing more things than one could before.

The wonder of having broken out of their continental isolation still awed Europeans in the eighteenth century. The economist Adam Smith, not given to hyperbole, called the discovery of America and the passage to the East Indies by way of Cape Horn "the two greatest and most important events recorded in the history of mankind."[1] They prompted a new group, called naturalists, to investigate plants, rock formations, and animals, even insects, discriminating, explaining, organizing, categorizing, and conceptualizing the fruits of two preceding centuries of collecting.

Objective knowledge became the great desideratum, and it was to be realized through a new intellectual regime that followed a specified route. First came a hunch, or testable proposition, followed by skepticism and scrutiny, passing then to the careful design of experiments that could be replicated. Reasoning would follow an inductive line, going from the particular to the general through hypotheses, tests, analyses of results, and tentative conclusions in slow, deliberate steps. Published findings advertised success to a widening circle of avid readers. Unlike translating texts, conducting experiments was costly, so collaboration (or working with servants, if one was wealthy) became almost an imperative.

Specimens from the Western Hemisphere awakened a keen interest in nature's bounty everywhere, prompting the curious to turn their attention to the ordinary objects around their homes, while the rich displayed their wealth through extravagant collections of exotic items.

No puzzle produced more disputes than that of spontaneous generation. Anyone assaulted in summertime by fruit flies on their ripening peaches can grasp the problem. Where did the flies come from? How had their seeds or germs or larvae been introduced into the fruit basket? Aristotle thought that living forms could spring from nonliving matter, that they issued spontaneously from the "life force" in the corrupted material in which they dwelt. Surgeons confronted with infections in their postoperative patients believed that germs generated spontaneously within the body. A seventeenth-century recipe for creating mice seriously advised mingling sweaty underwear and the husks of wheat in an open-mouthed jar and then waiting three weeks. Maggots were believed to generate on their own, as they seemed to breed quickly and from out of nowhere. Francesco Redi, an Italian physician, had a hunch that they actually came from eggs laid by flies. In what was one of the first controlled experiments, he put meat in various open, partially open, and sealed containers and discovered that maggots appeared only in the open ones. A half century later, John Tuberville Needham, a Catholic priest attracted to the same question, boiled tainted wheat and meat broth in separate flasks, which he then sealed. When microbes appeared a few days later it gave him confidence that organisms could be generated spontaneously. Within the clubby world of natural philosophers news of his experiments spread rapidly.[2]

The Dutch physician Jan Swammerdam became a fervent critic of the idea of spontaneous generation. As noted, he had used his skills in dissecting tissue to give close attention to the habits of mosquitoes, butterflies, wasps, snails, dragonflies, and bees. Swammerdam laid the foundation for much later work in entomology. Working in the 1660s and '70s, he studied the life forms that sprang from pond slime and rotting meat in order to lay to rest Aristotle's ideas. His careful inspection of the internal organs of insects revealed that the egg, larva, pupa, and adult were but different life stages of the same insect.

Another Catholic priest, the Italian Lazzaro Spallanzani, who lived in the mid-seventeenth century, liked solving problems, so he took on Needham's "spontaneous microbes." Spallanzani pursued biological experiments from his university position as a professor of logic and metaphysics. His cousin, the redoubtable Laura Bassi, the first woman with a regular position in a European college, gave him his scientific education. Her great accomplishment was introducing Newtonian physics to Italy, but her young protégé roamed widely over modern and ancient languages, mathematics, and natural philosophy.

Endowed with as much charm as curiosity, popular as a lecturer, and lionized by his contemporaries, Spallanzani studied digestion, the circulation of the blood, fertilization in animals, the senses of bats, and the regeneration of legs in salamanders. He pioneered the in vitro fertilization of a frog and artificial insemination of a dog. He also demonstrated that digestion was a chemical process of dissolving food, not, as had been thought, a mechanical one of chopping it up.

Spallanzani's work on spontaneous generation began by repeating Needham's experiments. He dropped seeds, peas, and almonds into a dozen or so flasks. Some he boiled for over an hour and then sealed tightly; another group he boiled for a few minutes and sealed, while a final batch he boiled briefly and just corked. After days, he opened them all and found that the sealed flasks that had been boiled for over an hour had no microbes inside; those that had been boiled briefly and sealed had a few; and the briefly boiled, corked ones were full of living microbes. Spallanzani's experiment not only impugned the idea of spontaneous generation; it also showed that some microbes could live after being boiled, if still exposed to air. As he put it, all living things have parents, but they are often invisible.[3]

Needham did not give up so easily and kept on writing about "a mysterious something" and "the great Vegetative Force" that produces things spontaneously and accounts for Eve's growing from Adam's rib. Spallanzani continued boiling organic matter in flasks, coming out with the same results.[4] The argument persisted until the following century, when the French Nobel laureate Louis Pasteur put the speculation to rest with his work on sterilization. The French Academy offered a prize in 1859 to anyone who could definitively resolve the conflict. Pasteur not only showed that microorganisms could not grow spontaneously, he also proved their abundance everywhere, including the air. As for the fruit flies on our peaches, we now know that the flies can smell fruit from long distances and are small enough to squeeze through most window screens or door jambs.

The endurance of the controversy points to a central problem of empiricism: the difficulty of replicating experiments. The Baconian emphasis on building theories inductively from facts gathered through observation and experimentation had triumphed in the seventeenth century, but not without introducing problems of its own. Without protocols strictly defined and adhered to, experiments could come up with as many results as experimenters, as in the case of Needham and Spallanzani. This led to a demand for better instruments and a better means for preserving and communicating results. Artists specializing in botanical and zoological drawings became more precise, and they pressured printers for better ways to reproduce their work. Most naturalists preferred to accept anomalies rather than rush to untenable generalization. Descartes was not alone in trusting none but his own observations.[5]

One physicist, trying to infuse his students with discipline,

said that an observer "must have unwearying patience, an attention that no circumstance escapes, a prompt and lively penetra-
tion, a wide and moderate imagination, a great deal of caution and
circumspection in his judgments."[6] R.A.F. de Réaumur, working
in the early eighteenth century, showed that even with exceptional
talents some things were hard to see, including how male frogs
fertilized the eggs of females. When female frogs are ready to lay
their eggs, male frogs leap on them and lock their bodies on top,
remaining there for days on end. Exactly how semen entered the
females was obscure. In the study of his country estate, Réaumur
placed pairs of mating frogs inside bell jars, in hopes that thus
captured, they might reveal what the males emitted. Using the
imagination the French physicist recommended, Réaumur had
little pants made for the male frogs. Waxed taffeta proved to be
the best fabric. Trial and error taught the optimal size and timing
for suiting up. Despite repeating these experiments over and over
for several springs, neither he nor his assistants ever caught the
male in the act of releasing its semen.[7]

　　Réaumur's repeated observations of mating frogs, like Spallanzani's disproof of Needham's "spontaneous microbes," represented investments of thousands of hours. Locating the sexual
organs of creatures and observing their operation were critical
achievements in the debate about spontaneous generation. When
Spallanzani boasted that he had examined 2,027 copulating frogs
and toads, he was also revealing the great human effort involved
to get nature to yield its secrets.[8] Yet the stakes were high. Investigations, when tempered by patience, could pry open some of the
most vexing perplexities in the world of nature.

　　Spallanzani became interested in whether animals such as
newts could regenerate an injured part of their body. He turned
to the drawings that Bernardo de Sahagún, the Spanish Franciscan missionary, had commissioned when chronicling the customs
of the Aztecs. In them was a description of axolotls, the stunning

little amphibians capable of regrowing a limb. It flourished in the now diminished freshwater lakes around Mexico City. Sahagún brought the axolotl to European attention. With it, Spallanzani demonstrated cellular regeneration. The scientific community has never lost touch with the axolotls, which now thrive in laboratories doing stem-cell research.[9]

The implications of work on spontaneous generation went directly to questions about the origins of human societies. Hippocrates and other classical thinkers had explained differences in appearance and culture among societies through the effect of climate and geography. Christians believed that all human beings were descended from the sons of Noah, who had been dispersed over the face of the earth after the flood. Columbus's voyages provoked rampant speculation on the origins of the Amerindians, but the lack of agreement invited the conjecture that perhaps the people of the Western Hemisphere were not from any place known to the Bible-reading Europeans. What if human variety did not represent modifications of Noah's lineage, but rather a separate line of creation?

The Amerindians had set off a search among Europeans for explanations of human differences that the awareness of African and Asian people had not.[10] An even more subversive idea occurred in the next century. If there were a mechanism for spontaneous generation, might the concept of a creator of all species have to be abandoned? The word "indigenous," which Europeans used to designate Amerindians, suggested that the natives had sprung from their particular setting and were not descendants of the survivors of Noah's Ark. The issue of spontaneous generation among fruit flies and microbes proved easier to resolve than that of human origins.

Isaac La Peyrère made a rigorous study of accounts of human origins in Genesis. His *Men Before Adam*, published in 1656, raised a storm of protest. He conjectured that there were two moments of creation, one much earlier for gentiles than Adam's line for Jews. In his interpretation gentiles were those outside the Judeo-Christian communities. His writings laid the basis for the concept of race, a term that originated with animal husbandry. It solved some problems of logic by providing a source for the wife of Cain as well as evidence that many of the earth's civilizations were much older than the time allotted in Genesis. Still, the idea of polygenesis provoked passionate opposition. If all men and women were not the progeny of Adam and Eve, as the monogenetic theory asserted, what happened to the unity of mankind? Without the presumption of consanguinity, what hope was there for peace on earth under a common God? Objections to polygenesis led to further speculation challenging the uneasy consensus resting on the mingling of ideas emanating from Athens and Jerusalem.

The outpouring of observations coming from the studies, libraries, philosophical societies, and museums strewn across Europe made it increasingly difficult to keep track of natural phenomena, much less make sense of them. The Greeks had been master synthesizers, arranging what they knew about the natural world according to the form and qualities of specimens, but findings from the New World had shaken confidence in the famous Greek systems promulgated by Aristotle, Ptolemy, and Pliny. Still, the ancients had left Europeans with an expectation of order existing among known phenomena. Order also suggested uniformity. It might have been possible to accept diversity, but eighteenth-century naturalists looked for the universal aspects of plants, animals, and human beings. In this choice—and it was a choice—a deep cultural trait of Western thought manifested itself in a preference for an underlying, all-encompassing order.

Experts developed a heightened sense of the need for more rigorous methods to deal with natural phenomena. Someone was bound to come up with a system for sorting things out, and the man who did—Carl Linnaeus—was one of the odder characters in the history of scientific inquiry.[11] Linnaeus's father had hoped that his son would follow him, his grandfather, great-grandfather, and great-great-grandfather into the ministry. With this expectation, he taught young Carl Latin at the same time that Carl was learning his native Swedish, giving his son a linguistic foundation for establishing a universal code for naming plants.

When Linnaeus rejected the ministry as a calling, his father was ready to apprentice him to a shoemaker. Here entered the first of many sponsors whom Linnaeus attracted in his youth. Recoiling at the thought of a talented young man making shoes all his life, a physician in Linnaeus's hometown in an impoverished northern province pushed him toward medicine as a means of earning a living. Fortunately for Linnaeus, medical education then encompassed botany because of doctors' reliance upon medicinal plants. With an impressive memory, an outsized curiosity, and a precocious interest in flowers, Linnaeus was primed to bring order to the proliferating knowledge about the natural world.

From the age of twenty to twenty-eight Linnaeus shuttled between penury and accomplishment as he pursued his education. At Uppsala, his aging botany professor hired him to give his lectures. Linnaeus next got permission to remodel the university garden. A small treatise he wrote on the sexes of plants in 1729 attracted the attention of another Uppsala professor of botany who sponsored his appointment as an adjunct.

These men who saw so much promise in Linnaeus pointed him in the right direction. They helped sustain him as he fashioned a career in science, or, more specifically, the taxonomy of plants, animals, and minerals. His patrons gave him more than encouragement and money; they also allowed him to plough through their botanical book collections, a critical privilege in a day when books were rare and access to private libraries even rarer.

The Uppsala Science Society sent Linnaeus, just turning twenty-five, to Lapland to study its flora and fauna. He was encouraged to report any other intriguing things he might find. Many of Linnaeus's idiosyncratic qualities can be viewed through the lens of this excursion. He covered an amazing 2,000 miles by foot and horseback in eighteen days, but he reported covering 4,600 miles to the society that was paying his expenses by the mile. He relied upon the local lore and guidance of the Lapland-ers, but he reduced their role in his account of the expedition to that of innocent bystanders. He wrote as though he had trekked alone through Lapland when in fact he had used guides and log-ging roads across the Scandinavian tundra.

Linnaeus studied the Lapps, the reindeer they ate, and the lichen that the reindeer ate. He noted the techniques the Lapps had incorporated into their way of life. His frequent backward glances to the weeks he spent in Lapland later inspired various narrative inventions. Linnaeus had returned with an elaborate ceremonial robe complete with native regalia that he subse-quently wore to ameliorate the inferiority he felt in the company of distinguished Europeans. In retrospect, Linnaeus stretched the length of the excursion from its true eighteen days to months and, in one account, a year, but he stayed true to his goal of demon-strating the utility of botany to Sweden's economic policies.

Back in Uppsala, Linnaeus was still poor and still intent on a medical degree. He proceeded to the University of Harderwijk in the Netherlands, something of a mail-order outfit, whose moder-ate fees and low standards appealed to him. The whole of medi-cal science in the early eighteenth century, it has been said, could be grasped by a quick student in eight days. Linnaeus earned his degree in less than a fortnight with a thesis on the cause of inter-mittent fevers in malaria. In this short time he defended the the-sis, passed an oral exam, diagnosed a patient, and supervised the printing of his notes![12]

Before arriving in Harderwijk, Linnaeus stayed with the

mayor of Hamburg, a visit that foreshadowed his adamancy about accuracy—at least in natural phenomena. The mayor had a prized possession that he hoped to sell for a large sum. Offering it for examination to Linnaeus, he must have been appalled to learn that the jaws, feet, and skin of his seven-headed animal, supposedly representing the Beast of Revelation, had been taken from various real animals and glued together in a bit of fraudulent taxidermy.

Making his observations public, Linnaeus departed promptly for Harderwijk. His visit to Hamburg, his first outside his native Sweden, stunned and embarrassed him. The riches he found there made the relative backwardness of Uppsala a personal humiliation. Struggling in his diary to reconcile patriotic fervor with reality, Linnaeus adverted to the happy Lapps he had lived among as evidence of Sweden's moral superiority. When he visited Amsterdam, he began wearing his Lapp costume complete with a shamanist drum to impress his new acquaintances with what was special about Sweden.

Botany has the advantage over entomology of involving objects that could embellish the grounds of estates. This encouraged a number of amateur horticulturalists to create gardens where they could display rare botanical specimens. Their beauty suggested taste rather than the physicians' concentration on medicinal value. Typical of the wealthy patrons of botany, one Amsterdam banker, George Clifford, used his extensive business connections as a director of the Dutch East India Company to tap into the world's store of plants. Since Dutch trade encompassed the coasts of Africa, the Far East, and the Western Hemisphere, he was able to amass specimens from around the world. Invited to write a definitive description of the Clifford collection, Linnaeus spent a year as its botanical curator with permission to buy any book or plant that he wanted.

Leaving the Low Countries with degree in hand, Linnaeus traveled through France and England, making a kind of pilgrimage to notable figures in natural history, most of whom gave him a generous welcome. When in Leiden he showed his host, Jan Frederik Gronovius, a manuscript dealing with a system for organizing the plant world. Gronovius was so astounded by its quality that he forthwith funded its publication. And thus the *Systema naturae* (*General System of Nature*) came to be published. It earned the twenty-eight-year-old Dr. Linnaeus an impressive reputation, which he vindicated during the next half century with a steady stream of publications and public lectures.

Once back in Sweden, Linnaeus married and then established himself in Stockholm as a specialist in treating syphilis. Three years later he joined the Uppsala University faculty as a professor. His lucrative practice with the Swedish aristocracy, including the royal family, intensified his contempt for the idle rich. Still, his practice gave him the money he needed to devote himself to his passion: the creation of a universal system for designating and describing plants, animals, and minerals. Such a system, Linnaeus was certain, would clearly reveal the hand of God in nature. Nor did he neglect his promotion of Swedish self-sufficiency, undaunted by the country's limited growing season. He never left home again.

The challenge in plant taxonomy was to devise a classification system that was simple, precise, and comprehensive. Linnaeus got his central concept from a small volume written by Sébastien Vaillant, a French botanist who worked at Paris's Jardin des Plantes in the early eighteenth century. Vaillant's pamphlet opened Linnaeus's eyes to the importance of the stamen and pistil in the reproduction of plants. Linnaeus had already come to the conclusion that plants and animals shared anatomical characteristics, diseases, and patterns of hibernation, so the idea that sexual organs were as important in plants as in animals didn't seem far-fetched to him. Henceforth, rather than look at all aspects of a plant, he focused on plants' reproductive organs.

The overtly sexual basis of Linnaeus's taxonomy both excited and disturbed his contemporaries. Linnaeus seemed to revel in it, going so far as to describe the leaves of flowers as bridal beds. The creator, he waxed eloquently, had perfumed the "noble bed curtains . . . with so many soft scents that the bridegroom with his bride might there celebrate their nuptials. . . ." He even classified plants that did not have obvious sex organs as those "with a hidden marriage." One outraged scholar denounced the Linnaean system as "loathsome harlotry." If it is true that one should not, as Mark Twain noted, make an enemy of a man who buys his ink in gallons, it is even truer that one should not provoke someone who is in the business of naming the members of the plant kingdom. Linnaeus named a common weed—the siegesbeckia—for an unlucky critic, Johan Siegesbeck.[13]

Now Linnaeus had to get others to accept his system, for its utility depended upon its widespread use. He also had to make sure that all newly discovered specimens could be fit into their appropriate place in the nomenclature. Continuing to perfect his taxonomy, in the 1750s Linnaeus worked out the binomial system of genera and species still used today. Never one to hide his light under a bushel, he expressed the conviction that he had opened up "a new epoch," going so far as to say that his science provided "the light that will lead the people who wander in darkness."[14]

Despite such exaggerations, there was nothing remarkably original about the choices Linnaeus made to bring order to his subject. His taxonomy triumphed because of its clarity and the diligence with which Linnaeus collected and catalogued specimens. He was a determined and adept promoter. In happier examples than siegesbeckia, he named plants after famous contemporaries, hoping to entice Europe's most important botanists and zoologists into adopting his classification system. Surnames were Latinized for both the genera and species, as with siegesbeckia or, more happily, the gardenia named after Alexander

Garden, the Scottish-born botanist in the colony of South Carolina who maintained an active correspondence with naturalists in
Sweden and Great Britain.

Linnaeus started with a hierarchy for all living things, including human beings. First came the kingdom, then phylum, class,
order, family, genus, and species, a ranking captured by thousands
of students in the mnemonic sentence, "Kings play chess on Fri-
days, generally speaking." The last two categories, genus and species, make up the identifying binomial that Linnaeus established
in the 1750s. He followed a strict protocol, capitalizing the genus
and writing the species in lowercase, as in *Convolvulus tricolor*, the
official name for the morning glory. He always used Latin or a
kind of Latinizing for his nomenclature, so, for example, the great
families of the rose became Rosaceae and of the lily Liliaceae.

Despite Linnaeus's emphasis on order and consistency, nature
didn't always cooperate. The ginkgo tree, one of the few specimens to have survived the last ice age, is a genus with only one species. In fact *Ginkgo biloba* is phylum, class, order, genus, and species
all by itself since other ginkgo species disappeared three million
years ago. On the other hand, the colorful family of asters, daisies,
and sunflowers—the Compositae—contains 22,700 species spread
across as many as 1,620 genera. Linnaeus's slim eleven-page pamphlet introducing his taxonomy grew into a multivolume work
identifying close to eight thousand genera by the time of his death.

Fittingly, Linnaeus's last name came from a big linden tree
growing at his family home. In the early eighteenth century Swedes
began to choose last names, instead of following the patronymic
system of using their father's first names as a last name, as in the
composer Eduard Larsson or Kristin Lavransdatter, the heroine
of Nobel laureate Sigrid Undset's trilogy. Linnaeus's father was
Nils Ingemarsson after his father, Ingemar, but he decided to
latinize the linden tree to give his family an unvarying last name,
perhaps foreshadowing his son's career of naming things.

The most famous item in the Linnaean system, of course, was *Homo sapiens* of the Animalia kingdom, Vertebrata phylum, Mammalia class, Primates order, and family Hominidae. Linnaeus himself liked to compare his categories to an administrative map of kingdom, province, territory, parish, and village. The term Mammalia, literally "of the breast," was his invention. No rubric, whether mammals, quadrupeds, or oviparous (bearing young alive), covered the full range of animals in the group, so Linnaeus had a choice to make in naming his own class, and that choice bears inspection. It excludes the male half of the category and focuses upon women as a physician might, especially one living when legislators were busy promoting pregnancies and breast-feeding.[15]

A prolific if not particularly popular educator, Linnaeus sent dozens of students around the globe searching for plants and animals that had eluded categorization. With full instructions of where to go and what to look for, Linnaeus dispatched his students to North and South America, Antarctica, Australia, Japan, Siberia, China, the Spice Islands, the Middle East, and the east and west coasts of Africa. Assigned to locate useful plants, they diligently sent back thousands of herbarium sheets mounted with dried plant specimens to use for identification.

As with the contemporary medical education, preparation for doctorates in botany was brief. Between 1741 and 1776, 186 graduate students under Linnaeus's direction emerged from Uppsala with their doctorates. He often wrote their dissertations. But if their studies were not rigorous, their postgraduate dedication often was. Because of their exertions, Linnaeus was able to examine 7,700 plants and nearly 4,400 animals by the end of his life. Five of his students perished collecting specimens, a record overtaken a generation later when eight zoologists working in the East Indies succumbed to tropical diseases.[16]

Even more assiduously Linnaeus cultivated knowledge of Swedish plants. In one of his many ingenious projects, he sent students out to the countryside to follow dogs, pigs, goats, horses, and sheep all day to see what fodder plants they would find if left free. This rather bizarre exercise was intrinsic to one of Linnaeus's major goals: to use botany to aid the Swedish economy. Fear of famine never left his mind, engraved there by periods of hunger he had witnessed in his childhood. He chided his fellow Swedes for ignoring their natural surroundings in their desire, as he said, to "eat like an Englishman, drink like a German, dress like a Frenchman . . . and guzzle vodka like a Russian."[17]

Linnaeus said that "no science in the world is more elevated, more important and beneficial, than economics, since all people's material welfare depends upon it," but he gave the subject an idiosyncratic definition as the study of practical uses of natural resources. The Royal Swedish Academy of Sciences followed his lead with the stated intention to publish "new remarks, inventions, discoveries, and experiments, which will serve the growth and development of useful Sciences, Economy, Trade, Manufactures, and several publicly necessary Arts and Artisanal trades."[18]

Linnaeus believed that he could build a scientific wall against future starvation by canvassing all the available edibles. If Sweden had neither colonies nor European extensions, it could bring the world's wealth within its border by extending the range of plants to be harvested as food or fodder. Harking back to an old-fashioned mentality that stressed hoarding provisions, he wanted to save Sweden's gold by growing goods at home instead of importing them. He even longed to plant mulberry trees to stanch the flow of Swedish gold spent on foreign silk. (A generation later, Jefferson tried to do that in Virginia.)

Even more important than the economy to Linnaeus was religion. His strict Lutheran principles mingled inside him with a messianic conviction of his own destiny. He believed that God could be approached through the study of Nature with a capital

"N." More pertinent, he considered himself chosen by God to reveal the wonders of the created universe. He saw no incompatibility between science and religion, because he had faith that more knowledge would only increase awe at the powers of the creator.

Still Linnaeus was not sentimental. When the Lutheran archbishop of Uppsala chided him for placing humankind among other primates, Linnaeus replied airily that he knew of no way that would follow from the principles of natural history to make a generic difference between humans and simians. Either the archbishop should find one or cease his complaints. He was certainly not going to let an archbishop challenge his understanding of God's universe.

Linnaeus's accomplishments appealed to a wide range of eighteenth-century luminaries. People far more sophisticated than he were drawn to his cataloging of the globe's bounty. Jean-Jacques Rousseau claimed that he had learned more from Linnaeus's *Philosophia botanica* than from all other books on morality. He knew of "no greater man on earth." Goethe claimed that no one had influenced him more than Linnaeus, excepting Spinoza and Shakespeare. His compatriot, the playwright August Strindberg, called Linnaeus "a poet who happened to become a naturalist." Over time he became "the Pliny of the North." Linnaeus shared these opinions of himself, announcing that "no one has more completely changed a whole science and initiated a new epoch," and "no one has become more of a household name throughout the world." Modesty did not figure among Linnaeus's many virtues. As he liked to say, "God created; Linnaeus organized."[19]

The cover of some editions of *Systema naturae* depicted Linnaeus in the Garden of Eden, a location critical to his project of naming plants because he shared the belief of some of his countrymen that Sweden was the oldest country in the world and that due to its remote place on the globe, it had escaped the linguistic disaster of the Tower of Babel. Accordingly, this made Swedish the language that Adam and Eve spoke! This conviction led inexorably to Linnaeus's view that he was restoring original names and thus collapsing the distinction between name and thing. Still, his desire to reach beyond a Swedish-speaking group led him to use Latin.

Despite his lifetime of work analyzing the features of plant life, Linnaeus never arrived at a satisfactory definition of a species. It slipped into common usage without a clear reference. He vacillated on the question of whether species changed over time; he was reluctant to conclude that they were not stable. What troubled him more was why the Almighty would have created such diversity and proliferation of species.

These were the very questions that had begun to bother experts toward the end of Linnaeus's career. Indeed, it was largely in response to his taxonomy that the questions became insistent. Linnaeus did have the dubious distinction of introducing a hierarchy of races to La Peyrère's earlier concept of race. It turned out to be an easy step to take for a man given to order and authority. In the second edition of Linnaeus's *General System of Nature* of 1740, he designated four groupings of human beings, noting that the skin color of each was confined to one of the four known continents. Perhaps it was an effort to spare God the embarrassment of proliferation that led Linnaeus to reduce the many shadings of human skin color to four. Overlaying this taxonomy onto geography also implied a divine intention, if not an actual cause.[20]

The English cleric John Ray flourished a generation before Linnaeus. He had developed a classification of his own that was notable for taking into account all parts of a plant—flowers, seeds, fruits, and roots—throughout the organism's life cycle. Like Linnaeus, whom he influenced, Ray had been an avid collector of plants, animals, and rocks as a child. Moving beyond the amateur's delight in perceiving the physical characteristics of a flower or shrub, he studied their forms in detail. Ray experimented with plant physiology and showed how living trees conduct water through their trunk and branches. Ray did more than report experiments; he brought some order to the confusion of names in use. Both Ray and Linnaeus emphasized the sharp distinctions among species. Neither could imagine one species becoming another. These were questions for their intellectual heirs.

Ray's books covered both theology and natural history. As deeply religious as Linnaeus, Ray expressed with great clarity his conviction that studying nature brought people close to God, something of an eccentric view when he published *The Wisdom of God* in 1691. Fossils also attracted Ray. His Christian faith prevented his concluding that God would let any of his creatures become extinct, but Ray did insist that fossils represented the remains of life forms. He rejected the conjectures enjoying some currency among his contemporaries that fossils were a sport of nature or a model that God had rejected or deposits left by a cunning devil. William Paley, author of the popular nineteenth-century work *Natural Theology*, which Charles Darwin's father insisted that he read, relied heavily upon Ray's *The Wisdom of God*.

Having a lively religious sensibility and a pulpit from which to express it, Ray revived the biblical idea that the study of nature is an appropriate form of worship. A close observer of the relation of forms to functions in living things, Ray interpreted the diverse ways that plants adapted to different environments as evidence of divinity.[21]

As Linnaeus demonstrated, a career in medicine was a great fit with botany. Hans Sloan earlier had also managed to pursue his passion for botany while maintaining a lucrative practice among the nobility and royalty of Georgian England. Sloan went with the Duke of Albemarle to Jamaica in 1687, staying over a year to collect new species of plants, among them cocoa, which became a new item for apothecaries before candy-makers exploited its commercial value. Sloan was created a baronet, the first physician to be

so honored. His estate in the Chelsea section of London became
the setting for the Chelsea Physic Garden, where he assembled a
vast library. He filled his cabinet of curiosities with such items as
a whip made from the hide of a seacow, or manatee, that he had
picked up in Jamaica.[22] This collection formed the nucleus of the
British Museum.

Another medical man, José Celestino Mutis, left his native
Spain in 1760 to go to South America as the private physician of
the new viceroy of New Granada. Already fascinated by herbal
medicine, Mutis made a study of cinchona, the source of quinine that the Amerindians used to treat fevers. Seeing the great
scope for studying botany in the New World, he proposed a royal
expedition to collect specimens. Although he had to wait two
decades for its authorization, Mutis was able to spend the remaining twenty-five years of his life in an exploration of some 5,000
square miles encompassing tropical, plateau, and mountainous
areas. When Alexander Humboldt visited him in Bogotá in 1801,
Mutis had thirty artists who had been working for years painting
his twenty thousand specimens.

Not content with finding and identifying new plants, Mutis
harvested samples, appending detailed descriptions of each species and its uses. Meticulous as a collector, he sent back to Spain
105 boxes of dried plants, drawings, shells, resins, minerals, and
skins with pertinent maps and field notes. Alas, those on the
receiving end did not share his enthusiasm and carelessly stored
the boxes in a toolshed. Meanwhile, Mutis pursued his work in
mineralogy, astronomy, mathematics, and sociology while keeping
up with an active practice of medicine.

In these early botanists we can see a new, spiritualized wonder
at the created universe. As the influence of superstitions and belief
in witches was abating, there was a disenchantment with magic
and mystery and a reenchantment of nature, now accessed through
observation, experimentation, and study. Human beings—at least

those who were now called savants—adopted a Promethean attitude without evident fear of Prometheus's punishment.[23]

Four years after Linnaeus published the first iteration of his taxonomy in *Systema naturae*, Georges-Louis Leclerc, better known as the comte de Buffon, became curator of the Jardin des Plantes in Paris. Born in the same year, 1707, the two men could not have been more different and, as it turned out, more opposed to one another. Buffon was a self-assured aristocrat, Linnaeus a diffident commoner. Buffon loved wild animals, while the devout and practical Linnaeus concentrated on useful domestic plants. Buffon went on a grand tour of Europe to avoid expulsion from school for fighting a duel, a kind of behavior repulsive to Linnaeus. If not a bon vivant, Buffon certainly frequented fashionable salons. The study of nature turned Linnaeus to God; Buffon's theories got him in trouble with church authorities.

Unlike the stable state of species Linnaeus favored, Buffon developed a new appreciation of how animals and plants developed differently because of diverse geographic locations. Identification of species drove Linnaeus; advertising diversity, Buffon. If Linneaus entertained bizarre notions about the language spoken in the Garden of Eden, Buffon publicized extravagant speculations celebrating nature's plenitude. Yet, after years of considering nature's conundrums, they gravitated toward each other—or at least their conclusions did.

A lover of all things large, Buffon complained that it was necessary to carry a microscope to identify plants in the Linnaean classification system. Worse, the stem, the flower's shape, and the leaves had to be ignored in deference to the stamens, and, if one could not see them, Buffon complained, "one knows nothing, one has seen nothing." To him, the "nomenclaturs," with their insistence upon the distinctions necessary to classification, destroyed the unity of nature. More profoundly, Buffon came to see that the Linnaean system inhibited a more robust interest in nature.[24] Neither man had been tamed by scholarly models of behavior in their disputes. There's a reason that academic subjects today are called disciplines; they monitor procedures and findings, something foreign to the eighteenth-century naturalists.

While Linnaeus was sending students around the world in the service of his sexual system for identifying plants, Buffon was working on the thirty-five volumes of his *Histoire naturelle, générale et particulière*. The first volume appeared in 1749, the last nine after his death in 1788. By grouping animals by region, Buffon stimulated others to generalize about the influence of the animals' origins. In his criticism of Linnaeus he carried much of the French intellectual community with him, though obviously not Rousseau, who had heaped such extravagant praise on Linnaeus. The French engagement with nature in the second half of the eighteenth century can largely be attributed to Buffon's work, which nearly every educated European had dipped into.[25]

Being contentious as well as curious, Buffon entered the fray in other intellectual disputes. The most fraught of these dealt with the supremacy of mathematics in science. The work of Copernicus, Galileo, and Newton had established mathematics as the queen of the sciences, with astronomers and physicists her principal courtiers. Buffon rejected the Cartesian idea that only mathematics could provide a sound language for science. A mathematician himself, he asserted emphatically (he rarely spoke in any other tone) that living forms were far too complicated to be comprehended mathematically. Botany, zoology, and ethnology, he insisted, were fields not likely to yield to abstract formulations.

With great zeal, Buffon struck a blow at the arrogance of mathematicians. It would take more than one demonstration to dislodge this monarch, but Buffon deserves credit for beginning the effort to show that different methods were required for different inquiries.[26] It's hard to know if either Buffon or Linnaeus was familiar with the numerical sequences in flower petals and pinecones that the thirteenth-century Italian Fibonacci had found.

Buffon added greatly to the Jardin's collection and created the labyrinth that one can find there today. Considered by many the most prominent naturalist of his lifetime, he also launched a theory that tormented Thomas Jefferson. After examining specimens from the Western Hemisphere and reading the extensive travel literature about it, Buffon declared that life forms deteriorated in the Americas! In his words: "There is thus, in the combination of the elements and other physical causes, something antagonistic to the increase of living nature in this new world."[27] Miasmas, he said, diminished the New World's two-legged and four-legged inhabitants.

Expanding his argument, Buffon hypothesized that the damp and swampy environment explained why quadrupeds shrank. Only serpents and insects flourished there. Buffon had clearly read more travel literature from the tropics than from North America. When one of his contemporaries discussed Amerindians as lazy, unintelligent, gluttonous liars incapable of rational thought, he handed Buffon his bludgeon. Eyewitness accounts going back to Gonzalo Fernández de Oviedo gave him colorful anecdotes for painting his unflattering picture of New World creatures.

Unlike Linnaeus, who wanted to see God conserving his creative impulse, Buffon delighted in its exercise. The lamentable fate of humans and animals in the New World provided examples of change under the influence of the environment, another favorite topic of Buffon's. Some skeptics raised the question of how the varied climate of North and South America could have led to the uniform copper skin tone of the Amerindians.[28]

Buffon's disdain for American animals was almost comic. The count didn't content himself with droll observations; he went ahead and spun an entire theory of natural development from what he considered the distinctly inferior flora and fauna of the New World. The llama he considered a wretched version of the camel that the conquistadors first thought it to be. Anticipating the Cowardly Lion of *The Wizard of Oz*, he described the puma as a poor example of a ferocious animal—no mane and definite cowardly traits. The absence of elephants, rhinoceroses, hippopotamuses, and giraffes was telling. The only European animal to thrive in the New World, he announced, was the pig. Looking for an explanation for the swelling population of toads, frogs, and snakes, Buffon credited spontaneous generation, linking the New World's abundant worms and vipers with its putrefying puddles of sodden earth.[29] He evidently had not heard of the experiment that Spallanzani conducted in the 1760s.

These claims incensed Jefferson, who was minister to France when Buffon's fame was at its zenith. Returning to America, Jefferson sent a bulky package to the Jardin stuffed with the remains of a large moose and several elk and deer as proof positive that larger animals could be found in North America. Buffon wrote a thank-you note and promised to include the information in his next volume, but died before he could do so.

A Dutch geographer and diplomat, Cornelius de Pauw, then picked up Buffon's notion and wrote a whole volume on the inferior size of American natives. Where de Pauw differed from Buffon was in studying with exquisite thoroughness the indigenous population. He made himself the leading authority on the Amerindians, though he relied on the firsthand accounts of Vespucci, Oviedo, and others. His damning verdict focused on "the timidity of his soul, the weakness of his intellect, the necessary of providing for his subsistence, the powers of superstition, the influences of climate," which he concluded "all lead him far wide of the possibility of improvement."[30]

Even Europeans degenerated in America, de Pauw claimed, instancing the lack of a single book originating from Mexico or Peru. He drew heavily on the *Commentarios Reales de los Incas* of Garcilaso de la Vega, whose mother was an Inca—but perhaps that didn't count since the book had been published in Spain. De Pauw's dire conclusions, published in the 1770s, were overtaken by the outbreak of the American Revolution, which turned such North American creoles as Benjamin Franklin, George Washington, and Jefferson, all very tall men, into Parisian celebrities. The great nineteenth-century zoologist Georges Cuvier pointed to these strictures about size as the prime error of his predecessors.[31]

Consistent with his criticism of the Linnaean system, Buffon disdained microscopic studies. He obviously believed in the Latin motto *aquila non capit muscos* (eagles don't catch flies), saying that only lesser men should occupy themselves with little things. He himself was large, built more like a field marshal than a scientist, David Hume judged. Bristling with opinions, Buffon wrote his preferences for big, wild animals into his great masterwork.

Lots of hidden assumptions resided in Buffon's speculations. Species did not change. They might improve; they certainly could decline, but they didn't vary substantially in form. He also maintained that unvarying species were superior to variable ones. Yet Buffon's theory broke new ground by insisting that there were such things as species and defining them crisply as animals that could reproduce their own kind. And he wrote about all this with sparkling rhetoric. Linnaeus disliked Buffon because he saw him as a spoiled aristocrat. Buffon's disdain for Linnaeus went deeper.

In seeking to classify natural objects, Linnaeus stripped them of the aspects that fascinated Buffon, who wanted to study their habits, temperaments, and instincts. He speculated about the impact of place upon animals and even entertained for a while the idea of a connection between apes and humans. What Buffon failed to do was unlock the enigma that there were species with undeniable affinities and equally undeniable individual traits.

Considered a great stylist, Buffon wrote tomes filled with engaging essays about natural phenomena that competed for readers with Voltaire and Montesquieu. What he really produced was an encyclopedia of natural phenomena, each with its own colorful rendering. Ranging over all fields, he proposed that the geological formation of the earth came in stages, opening up the new field of paleontology. He further suggested that the planets might have been formed from a collision between the sun and a comet.

The ferocity with which Buffon defended his many insights won him enthusiastic support among the learned elite of Paris. They would rather have gone wrong with one of their own than be right with a bland foreigner. Buffon's boldness meant that he published whatever he thought, thus liberating his generation from many religious strictures. When he said that the world was seventy-five thousand years old, not the church's figure of four thousand, Sorbonne theologians reproved him. He then backed down to keep peace and confined his revised estimate of the earth's age at half a million years to the safety of his notes.[32] This alone was a major conceptual breakthrough in understanding the globe and its inhabitants.

By the eighteenth century most naturalists had become experts, but they weren't the specialized professionals of the nineteenth-century research laboratory. If they were on university faculties, they usually taught a variety of subjects. They might also practice law, run a business, or perhaps be gentlemen, like Réaumur, with enough leisure and money to indulge their scientific interests. They spread themselves widely among topics like botany, anatomy, zoology, geology, and psychology.

For all their brilliance, these gifted amateurs thought in a different intellectual milieu. We want to rush them into becoming scientists so we can better fit them into our own categories, but theirs was an age filled with more questions than answers and, hence, open to propositions that now seem ridiculous. Alexander Dalrymple, for instance, the Scottish geographer with the British Admiralty, was convinced that there must be a continent in the southern sphere as large as the landmasses in the north to "maintain the equilibrium necessary for the Earth's motion." Botany-loving doctors were not averse to selling quack remedies, and the love of exotica became an end in itself for many.[33]

Europeans' knowledge of the world was not so great that scientific chores had to be divvied up, but it did call for a better ordering of phenomena. The sciences got their modern footings in the second half of the eighteenth century with the organization and categorization within fields. Questions that had initially been directed to the people, plants, and animals of the New World turned toward all of nature's domain. Comparisons between Old and New World specimens launched fruitful investigations. At the same time comparisons winnowed away unique qualities and privileged those qualities, forms, or patterns that were common among them. Botanists following Linnaeus could ignore the fragrance, color, and shape of a rose in their concentration on the stamen and pistil. In these ways, those who wanted to master nature's secrets of production and reproduction were making the move away from the lovers of nature whose taste ran to planting gardens and painting flowers.

CHAPTER SIX

THE TRUE SHAPE
OF THE EARTH

R. de Tournieres's engraving of a portrait of Pierre-Louis Moreau
de Maupertuis, the French savant and explorer who led the
expedition sent to measure the latitude at the North Pole, 1737.

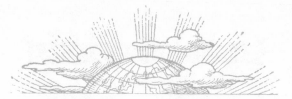

T he shape of the earth, the origins of human societies, the relationship that animals and plants bore to
one another, telling time at sea—even the presence of
gravity—divided experts in the eighteenth century. Settling scientific disputes and visiting the still pristine spaces remaining on
earth presented naturalists with a full agenda. As uncertainties
yielded to proofs and experts imposed order on the world around
them, contemporaries felt that they lived in an age of enlightenment. And their learned men did not fail them—a new voice
sounded in Europe's intellectual circles. It was lively, witty, and
exuded confidence in the possibility of commanding all extant
human knowledge.

A great many mysteries about natural phenomena remained to
be solved in the eighteenth century, and confusion still reigned
over the precise shape of the globe. It was so important that studying it had a word all its own—geodesy. Creating uncertainty about
something as basic as the shape of the earth made the ground
upon which Europeans stood no longer seem quite as solid as it
had before the shape became a contested issue. A dozen unsettled
questions swirled around the subject. Centuries earlier, Christopher Columbus had compared the earth to a pear except for the
stem "where it is higher, or as if someone had a very round ball,
and in one part of it a woman's nipple would be put there."[1] The
nipple, he believed, was located in Venezuela, where he had compared the mouth of the Orinoco with the Garden of Eden.

Passionate personal, philosophical, and national rivalries heated
up discussions about geodesy. In the 1730s the French Royal
Academy decided to tackle the problem. A significant portion of

its members—led by its distinguished Perpetual Secretary, Bernard le Bovier de Fontenelle—did not accept Newton's theory of gravity, which indicated a certain shape. These savants were prepared to spend good money proving him wrong.[2]

It takes a horse to beat a horse, and they found one in the great seventeenth-century philosopher René Descartes, who had offered a rival explanation for what kept the planets from running into each other. With his notion of movement as coming from push-pull mechanics, he rejected the idea of his contemporary, Johannes Kepler, that there could be action through a void to keep the planets on course. Instead, he conjectured, invisible matter actually filled the space we perceive as being empty, so that the planets didn't bump into the sun or each other because swirling around them were vortices of particles that deflected them into their own orbits.

Long dead when Newton put forth his theory about planetary movement at the end of the seventeenth century, Descartes would certainly not have liked the idea of gravitational attraction as a force in the universe. This was enough for Descartes's followers to drag him into the uncertainty about the shape of the globe, even though he never commented on the subject. Eighteenth-century Cartesians asserted that the earth was an elongated sphere like a Japanese eggplant, a somewhat tamer version of Columbus's pear. Newton maintained that the earth was more like a tomato—a spheroid flat at the poles, where gravity's action would be greatest, and bulging at the equator. The fact that Jupiter, the object of six years of careful inspection by Galileo, was flattened at the poles made it seem likely to Newtonians of the eighteenth century that this would be the case for the earth as well.

Some Frenchmen—most notably Voltaire—preferred Newton's idea of gravity. After spending three years in England, he came back home eager to change the minds of the learned men and women who gathered regularly in salons in private homes or in the biweekly sessions of the French Academy of Sciences. Jacques Cassini, who headed French mapmaking endeavors, held tenaciously to the view that the earth's shape resembled a thick tube rather than the Newtonian tomato, and he proved himself a tenacious adversary.

Such controversies presented interlocking theoretical and empirical questions, usually involving a knowledge of mathematics. In the rarefied atmosphere of the high-minded Parisian salons, elegantly dressed men and women argued about the shape of the earth. Pamphlets flew back and forth. With the academy's mandatory civility fraying, the astronomer Louis Godin proposed sending an expedition to measure the length of a degree of longitude at the equator. He committed himself and his cousin, Jean, to the effort. If a degree of longitude were greater there than farther north or south, then Newton would be proved right: the globe was more round than elongated. Hoping for a Cartesian outcome, the mathematician Jean Le Rond d'Alembert deemed it a "question of national honor not to let the Earth have a foreign shape, a figure imagined by an Englishman or a Dutchman."[3]

Two members of the French Academy, Pierre-Louis Moreau de Maupertuis and Charles de La Condamine, immediately threw themselves into Godin's plan to organize an expedition to Quito, near the equator, and, as a backup, to the North Pole. La Condamine and Maupertuis were not exactly the kind of men likely to head for the discomforts of the torrid tropics and the untracked snow of the Arctic Circle. Both were known for their wit and style, as well as their command of natural philosophy. They were greatly in demand in literary cafés, private gatherings, and even at King Louis XV's court.

Maupertuis studied mathematics so that he could read Newton's *Principia*. La Condamine used his mathematical skills in a rather different way. His scrutiny of the French lottery revealed that they inadvertently gave away more money than they took in.

He got Voltaire to put up the money for a scheme to game the system. The fortune they won cemented a closeness already forged by their shared admiration for Newton. When Maupertuis published a tract explaining Newtonian physics in accessible and conversational prose for a lay audience, Voltaire hailed him as "our Christopher Columbus."[4]

Both La Condamine and Maupertuis, lifelong friends, were among the sixty members of the prestigious French Academy where, according to Maupertuis, "there are people who are shocked simply by the word attraction."[5] Their sociability made them favorites. Maupertuis played the guitar. His first paper before the Academy dealt with philosophical reflections on musical instruments. The two men had a taste for adventure and neither was averse to taking risks. Maupertuis had given up his commission as a cavalry officer to devote himself to philosophical speculation. La Condamine, during his brief career in the army, demonstrated his insatiable curiosity. He set up a telescope on a hill to get a better view of the Spanish forces his unit was fighting. Only when his companions noticed that his scarlet jacket had attracted enemy fire did he come down from his commanding view.[6]

Fontenelle became rapturous at the prospect of sending academy members out to actually determine the shape of the earth. He had helped popularize Cartesian ideas through his graceful, conversational style of writing, but he was also delighted to push salon society toward natural philosophy. Evoking the hardships and perils of the expedition, Fontenelle predicted glory for "the new Argonauts." They would focus on measuring a degree of longitude for comparisons with the expected length for a perfect sphere. For this, Godin brought together a team of astronomers, mapmakers, and botanists. The botanists weren't expected to help with the astronomy; they were taking advantage of a royally funded research trip to gather plant specimens in the New World.

The academy succeeded in gaining permission from King Philip V of Spain to let foreigners enter his viceroyalty of Peru, where the measuring points in the province of Quito were located. The first in the Spanish Bourbon line, Philip, who was Louis XV's cousin, seized on the chance to burnish his reputation as an enlightened monarch and show that he had overcome the Spanish obsession with secrecy. Still, he did assign two young naval officers, Antonio de Ulloa and Jorge Juan, to keep tabs on what the French would be doing near Quito and Riobamba in present day Ecuador, an administrative center of his carefully guarded American empire.

No effort was spared by the organizers of the expedition. When the French naval minister, whose assistance was vital, became indisposed, Maupertuis visited him, guitar in hand, and sang his way to securing full support. Pierre Bouguer, who had achieved early fame for his work on the design of ships and the best methods for observing the altitude of stars at sea, joined the planning group. With funds in hand, they obtained from London the most advanced equipment for accurate measurements—a painful concession to English superiority in instrument making. They also received some shocks. When La Condamine went to the French treasury to examine pieces of Incan artistry—insects and animals worked in solid gold—he learned that the treasury, the supposed guardians of the Peruvian artifacts, had melted them down to send to Cartagena as gold bars. Governments apparently are always impecunious.[7]

"The new Argonauts" were ready to set sail for South America from La Rochelle in May of 1735. A mathematical prodigy, Godin became the nominal leader because it had been his idea, but he, La Condamine, and Bouguer, the three principals, soon quarreled, leading to a split of authority and independent side ventures while in Peru. By sheer force of personality, La Condamine became more closely associated with this remarkable first international scientific expedition. He also wrote the most voluminously about his experiences when he returned home.

Despite their rivalries and eccentricities, the leaders managed to cooperate on their most important goal: measuring longitude at the equator. They also shared a commitment to precision that kept them at their task for six years. They repeatedly measured the difference in angles from two separate observation points of the same star as it passed across the sky. From this information they could calculate longitude. They started by marking off a distance of 70 miles and synchronizing pendulum clocks at both ends of this line. With the flash or sound of cannons they coordinated their timing to make sure they measured their angle to the star simultaneously. Fixing their telescopes on the same star, they then calculated their longitudinal separation by tracing the star's passage across the sky.

Arduous anywhere on the globe, the hazards of doing all this in the Andes were horrendous. Godin, La Condamine, and Bouguer had to climb mountains to altitudes of 10,000 feet, dragging heavy instruments along the narrow paths that linked Andean peaks by way of swaying rope bridges that their Amerindian guides made. Insects, infections, and disease assaulted them constantly, but they continued to take sightings and make calculations over and over again across hundreds of miles of the Andes. Not content with this single effort, La Condamine and Bouguer took advantage of lunar eclipses to make calculations that improved the maps of South America's west coast, while the expedition's botanist collected thousands of new plants.[8]

Meanwhile, back in Paris Maupertuis convinced the Academy that measuring longitude at the equator needed a strong contrast to clinch the debate about the earth's shape. He won approval for an expedition to Lapland that departed just about the time his colleagues in Quito were setting up their telescopes for a first measurement. The King of Sweden gave the expedition the latest maps of the area along with the talents of Anders Celsius, Linnaeus's associate at the University of Uppsala. Celsius, who later devised the temperature scale that carries his name, went to London to secure the best telescopes for their calculations.

The sixteen months that it took to measure the length of a degree along a polar meridian hardly compares with the six years those who went to the equator spent in the Andes, but it was no picnic either. Maupertuis's team had to ferry themselves and their equipment up and down mountains and by way of rivers whose cataracts forced them into lengthy portages. Like the La Condamine party, Maupertuis's also relied on the labor of the local people, in this case a contingent of Finnish peasants in the Swedish army. Testimony to the polyglot world of eighteenth-century Europe, the group even included a translator who spoke Finnish, Latin, Swedish, and French.

The party of eight astronomers, five servants, twenty-one soldiers, and the interpreter leveled forests to make sight lines. In summer they ventured into insect-infested swamps that even the natives avoided. The mosquitos were so ferocious that, despite the heat, the men wore heavy reindeer coats. If all this weren't bad enough, they accidentally started a forest fire that consumed one of their marking stations along with hundreds of acres of forest. The months of good light were short, so they worked frenetically to lay out their markers and check their instruments while making observations.[9]

Spending the next spring and summer rechecking calculations, the astronomers and their servants returned to France in August 1737. The next year Maupertuis published his *La figure de la terre,* which put the case for the shape of the globe as we know it today. The tomato won! Cassini did not concede the battle even though messages from the group at the equator confirmed Maupertuis's conclusion that Newton was right: the globe did bulge at its center and flatten at the poles. Couched as an adventure story,

La figure de la terre won Maupertuis a Europe-wide reputation and reignited the embers of the geodesy debate. Members hailed his presentation to the academy as a model of clarity and transparency. Voltaire was beside himself: "If your operations are worthy of Archimedes and your courage of Christopher Columbus your description of the snows of Torneá is worthy of Michelangelo."[10]

News of the successful polar expeditions dampened the spirits of the weary investigators in the Andes. Bogged down confirming their figures, they feared that all the glory had already been gleaned by Maupertuis. While he was being lionized in Paris, they were battling storms so fierce that their servants had to lash shut the entrance to their mountain hut at night. When funds ran low, La Condamine actually set up a store in Quito from which he sold personal belongings such as embroidered clothes, fine Holland shirts, cotton clothes, bedsheets, silk stockings, gloves, switchblades, needles, gunflints, emerald jewelry, books (including a history of the French Academy), and a diamond-encrusted cross of Saint Lazarus. What makes this mercantile effort as astounding as it is amusing is the testimony to the extent of the goods—plus extremely bulky equipment and the belongings that the team held on to—that had been carried from France to Santo Domingo, then across the Caribbean to Panama, then loaded onto boats to cross the Isthmus through the Chagres River, and reloaded for the voyage down the Pacific coast![11]

Word of Maupertuis's speedy expedition to Lapland made "the new Argonauts" at the equator doubt that the leaders of the Arctic expedition had been as excruciatingly precise as they were in their measurements, and they were right. Godin, La Condamine, and Bouguer also comforted themselves that they were studying other physical properties of the stunning world around them. When they were not lost in clouds on their mountain climbs, they experimented with the speed of sound and the effect of temperature on materials. They even came up with a standardized land measurement: the meter. On a side trip, La Condomine found the caoutchouc tree that produced the resin that Indians used for waterproofing, later called rubber. He also discovered an unusual metal that Indians used for their jewelry, later determined to be platinum. In June 1742, all three leaders were satisfied that they had tied down every calculation.[12]

Getting home proved almost as exhausting as lugging equipment through snowstorms on Andean peaks. Spanish officials refused to let Louis Godin leave the country, insisting that he spend seven years teaching math at the University of San Marcos in Lima. The expedition's botanist then proved too valuable as a physician in Potosí to be allowed to depart. La Condamine did not press his luck trying to get permission to leave from Spanish officials, nor had he had quite enough adventure. He elected to go home by way of French Guiana, which called for rafting down the Amazon, a trip of 3,000 miles. He had enough leisure time on the great river to contemplate the memoir he would write at the end of his historic journey. After six years spent in jungles and on mountaintops, La Condamine remained a Parisian salon member, highly conscious of the audience awaiting his return.

La Condamine mapped the river and plumbed its depths. He frequently stopped to learn how the Indians hunted game with blowguns and poison darts. Among the animals he listed in his copious notes were bloodsucking bats, toucans, turtles, crocodiles, tigers, monkeys, electric eels, porcupines, sloths, snakes, and boars. He enjoyed the hospitality offered at a few Jesuit missions. One missionary told him that the population along the Amazon was one-thousandth of what it had been when the Spanish arrived. Perhaps an exaggeration, but it is another reminder of the devastation that conquest wreaked upon the indigenous population.[13]

Maupertuis's *La figure de la terre* had included useful information for the royal officials who paid for the trip: exact measurements, with plates showing the chain of triangles made to measure a meridian near the North Pole for those who could follow the mathematical proof that the earth was indeed oblate, and enough colorful anecdotes about Lapland to please the general public. Thirteen years later, at the end of his expedition, La Condamine published three books to accomplish the same goals. *Journal du voyage fait par ordre du roi à l'équateur* was a diary of his ten years in the New World and all of his many accomplishments. In his *Relation abrégée d'un voyage fait dans l'Interieur de l'Amérique Méridionale,* he told of his exploration of the Amazon River basin; *Mesure des trois premiers degrés du méridien dans l'hémisphere austral* presented the scientific details of the expedition.[14]

Bouguer and the two Spanish naval officers who accompanied him also published their accounts of the journey. All of the books—well reviewed and widely discussed—sustained public attention for over a decade. Readers, naturalists, and statesmen talked about the shape of the earth and the contours of South America as a topic of vast intrinsic interest. By sheer courage and endurance, with the intelligent application of their talents, Godin, Bouguer, Maupertuis, and La Condamine had moved beyond their reputations as esteemed members of the academy to win renown as expanders of the universe of knowledge.

In a poignant coda to the equatorial expedition, the last member arrived back in France in 1773—twenty-eight years after its departure. He was Jean Godin, cousin of the astronomer. He had joined the group as a general helper, scouting marker locations and sending signals for the principal investigators. He learned enough about natural phenomena from the expedition's experts to teach at the College of Quito at the end of the expedition. In 1741 he married the fourteen-year-old daughter of a local notable. After eight years of marriage, he learned of his father's death and decided to go to French Guiana—a trip down the Amazon like La Condamine's—to make arrangements for his family to return to France. Alas, neither the Spanish nor the Portuguese authorities would allow him to pass through their territory to retrieve his wife. Godin was in a quandary. He refused to go to France alone, and he couldn't get back to his wife. One lonely year after another passed—twenty-three in fact—while he pursued a succession of desperate schemes to reunite his family.

La Condomine, once back in France, was able to persuade the Portuguese king to help the distraught couple. Eager to please the French, the king sent a naval unit to fetch Isabel Gramesón Godin, who had not received any word from Jean in all these years. When she learned that there was a boat waiting for her on the Amazon, she decided that she would meet it halfway. Her father and brothers did everything to dissuade her from taking such a dangerous trip. Unsuccessful in this effort, they joined her for what became a truly horrible journey. A succession of disasters killed all of her relatives, leaving her to wander alone through the snake-infested tropics for weeks.

People were more patient in those days. The ship captain sent to give Isabel passage to French Guiana waited for two years on the river. Two Indian couples rescued her and guided her to him. Jean described Isabel's truly remarkable survival and their joyous reunion in a lengthy letter to La Condomine, who received it as he was busy preparing a new edition of his *Relation abrégée*. Always eager to usher a dramatic narrative into print, he added Jean Godin's letter to his text, giving the public a thrilling romance to add to their memories of these two stunning expeditions.[15]

During this period, maps had gotten better and better, but mapmakers had to contend with the fact that the mariners upon whom they relied for information had great difficulty determining where they were at sea. Lots of boats plied the waters of the Atlantic, Pacific, and Indian Oceans, but they kept to known sea-lanes. Straying from them created great risks to life and cargo.

Today navigators have instruments that enable them to calculate both latitude and longitude. Without these, captains in the eighteenth century had to depend upon dead reckoning, which involved figuring one's location by estimating how far, in what direction, and at what speed one's ship had moved from port, a method that piled errors upon errors. Latitudes were relatively easy to figure out because they were parallel circles around the globe. A trained navigator with an astrolabe or quadrant could determine latitude by measuring the altitude of the North Star or the sun, if weather permitted.

Longitudes were altogether another matter. Determining them baffled the most skilled navigators. Since the lines, or meridians, bent to meet the North and South Poles, they were curved, making it extremely difficult to find an east-west position at sea. It wasn't even that easy on land because it required knowing the precise time when making astronomical sightings. In 1657 Christian Huygens developed a pendulum clock that relied on gravity, each swing precisely ticking off one second, which greatly improved the calculation of longitude on land.

When the French Academy of Sciences took up the task of making an accurate map of France, they were helped by Galileo's pioneering work that had turned the moons of Jupiter into a celestial timepiece through calculations involving triangulating from a baseline on earth. To achieve stability for the baseline for the French map, workers cut varnished wooden rods for the seven-mile distance between Paris and Fontainebleau. It took them two years to determine that a degree ran for 69.1 miles, a figure that still stands. Their findings led to surprising results with some cities relocated 100 miles away from their old cartographical positions. France, it seemed, had shrunk, causing Louis XIV, the project's funder, to exclaim that the effort had cost him "a major portion of my realm." Accuracy had its price.[16]

Determining longitude at sea remained a challenge. Sending more and more ships farther and farther from land made it imperative to improve astronomical tables along with the telescopes, thermometers, and barometers that aided navigators in measuring lunar positions and calculating tides.[17] Philip III of Spain offered a hefty lifetime pension to anyone who could devise a way to calculate longitude at sea, while Louis XIV established an observatory in Paris to which he lured famous scientists from all over Europe to address the same problem. The challenge had already intrigued such luminaries as Galileo, Tycho Brahe, Kepler, and Newton. Navigators knew that every four minutes of difference in local time equaled one degree of difference in longitude, but clocks in that day lost about fifteen minutes every twenty-four hours.

There were still no clocks capable of keeping time on the ocean. The urgency for such an instrument became acute as shipwrecks became more common. When the English lost an entire fleet just 20 miles off its coast on a foggy night in 1707, the flaws of dead reckoning became painfully conspicuous. The disaster had an even more piquant lesson to teach. The night before the accident, a common seaman approached the fleet's admiral, Cloudesley Shovell, to say that he had been keeping a record of the ship's position and believed that it was off course. The admiral responded to this news with alacrity. He charged the man with mutiny and had him hanged forthwith. The law was on the admiral's side. It quite explicitly forbade "subversive navigation by an inferior," confirming once more how officials discountenanced an inquiring spirit.

Within twenty-four hours the seaman's figures proved correct. All four ships ran directly into the Scilly Isles. The two thousand soldiers and sailors aboard perished. The admiral, one of two survivors, was murdered by a marauder on the beach, but perhaps not before he had time to contemplate his rectitude in punishing unauthorized ingenuity.[18] The Scilly Island catastrophe prompted Parliament to set up the Longitude Commission, with a very handsome bounty established for the person who could figure out a "practicable and useful" means of locating one's longitude at sea.

Astronomers set about working on what they called the "heavenly clock." By this they meant measuring the motion of the moon more accurately. They did eventually succeed in developing a better quadrant, but it demanded someone with superior mathematical skills to be on board.

A clockmaker named John Harrison came up with an idea for developing an accurate chronometer, but in their educated arrogance, the commissioners disdained the proposal of a simple mechanic. Harrison spent forty years designing and redesigning a time-telling mechanism that would work at sea. Its accuracy was tested when an astronomer, sent to Barbados to determine the longitude there by observation of Jupiter's satellites, showed that Harrison's chronometer had pegged the longitude with an error of less than 10 miles. The experts on the august Longitude Commission were still loath to have an uneducated artisan carry off the reward. George III had to intervene to secure the prize for Harrison. A special act of Parliament bestowed it upon him for an invention "that would carry the true time from the home port like an eternal flame to any remote corner of the world."[19]

Curiosity about the world—its shape, size, and content—had finally become fashionable. Voltaire, upon his return from London in 1735, noted how much styles had changed; people weren't writing verse because they were now studying geometry, especially the ladies, he said.[20] The grand debate between the Cartesians and Newtonians had spread beyond the learned societies. The puzzle that had driven witty salon buffs to the Arctic Circle and equatorial tropics had led to publications that delighted the whole reading public and no doubt turned many an adolescent toward natural philosophy.

A critical change had also taken place in reading habits around mid-century. While people before had read and reread the Bible and devotional material, men and women now had an array of things to read—journals, novels, travel narratives, political pamphlets, and religious tracts—which were usually read only once. Information and entertainment were taking over from piety and wisdom as the rewards of literacy. Extensive reading replaced the intensive kind of the past.[21]

In 1745 Denis Diderot and Jean Le Rond d'Alembert, two of France's most prominent savants, were invited to participate in a French translation of Ephraim Chambers's *Cyclopaedia*, a pioneering compendium of facts that came out in England in 1728. Looking over the text, they decided that the advance of learning had been too great to make it worthwhile to reissue an old work. Instead they would organize the talent of their generation and produce something entirely new.

"We watched," Diderot wrote, "as the materials expanded before our eyes, the nomenclature became obfuscated, objects were brought in with numerous names, instruments, machines, and maneuvers, multiplied beyond measure, and innumerable detours of an inextricable labyrinth became increasingly complex."[22] Six years later Diderot and d'Alembert began publishing the *Encyclopédie*, a magnificent collection of information on the arts and sciences and much else. It came to symbolize the age of the Enlightenment.

Wanting to do more than present an aggregation of knowledge, Diderot and d'Alembert prescribed to their readers how to approach learning, as well as their lives. They made their own daily routine of solitary study and intelligent conversation a model for others. Reflection should balance sociability. For the fashionable who frequented salons, soaking up information from the *Encyclopédie* filled a reservoir of commentary for future conversations. Those who didn't circulate in that rarefied company could learn what the salonists were talking about through reading the *Encyclopédie*.

Running to twenty-eight volumes with over three thousand illustrated plates, the grand publishing venture took twenty-one years to complete. The editors, writing thousands of the more than seventy-one thousand entries, also continued to turn out essays for the French press on one controversial subject after another. Diderot actually spent three months in prison for raising questions about God's existence.

Diderot and d'Alembert didn't privilege theory over fact or literature and science over practical techniques. They actually visited dozens of workshops to write on the useful arts. They had plates made detailing the production of everything from clocks to centrifuges. Within the richly bound covers of the *Encyclopédie* readers could find pieces from all the major intellectual figures of their day: Voltaire, Montesquieu, Condorcet, and Rousseau and the intrepid explorers Maupertuis and La Condamine. The literate public could read about Linnaeus's grand system for categorizing plants, Réaumur's frogs, Swammerdam's mosquitos, Spallanzani's newts, Sloan's Chelsea Physic Garden, and, of course, all the specimens brought back from Lapland and Quito. Probably as exotic for most Parisian readers were the hundreds of entries describing industrial processes that went on just a few miles from their homes. One of the aims of Chambers, whose encyclopedia had prompted d'Alembert and Diderot to start one of their own, had been to universalize the Newtonian explanation of gravity and through that the shape of the earth that the French Academy's two geodetic expeditions had confirmed.[23]

In retrospect, what seems remarkable about Diderot as a thinker was his responsiveness to natural phenomena. His essays probed the nature of life with a prescient sense of the importance of the environment. He reviewed the dispute about spontaneous generation. His musings were not original—indeed he was quite an avid reader of the principal naturalists of his time—but his writing style irradiated dense subjects with a brilliant wit.

By the end of his life Diderot was an atheist, believing in a totally materialistic world without a design. As he said, speaking of living entities, the organs produce needs and needs produce the organs. In his *Le rêve de d'Alembert* (The Dream of d'Alembert), he speculated on the unity of life-forms and their perpetual state of flux. "Species," he wrote, "are merely tendencies toward a common end peculiar to them." Without in a way suggesting evolution, or the actual changing of species over time, he pushed other thinkers in that direction in his characteristically provocative way.[24]

The French Academy's Perpetual Secretary Fontenelle was such a man, though he succeeded better than Diderot in staying out of prison. Fontenelle is one of those historic figures of much greater importance to his contemporaries than to posterity. Vital until his death a month before his hundredth birthday in 1757, he participated in most of the consequential debates of his century. In his youth he took the side of the moderns in a literary war waged by the "Ancients," the academicians who maintained that classical taste and knowledge would never be surpassed. This position almost cost him membership in the academy since the Ancients succeeded in blocking his election four times.

Disinclined to make enemies, Fontenelle wrote about peoples and their religions with circumspection. He explored the universal human situation, which called forth ideas of a divine being. "Men saw clearly things that they could not produce," he wrote, citing lightning, wind, and storms. All people went through this stage, he argued, sometimes believing in many gods. In juxtaposing the myths and fables of the ancients with those of the Amerindians, he used comparisons in a fresh and powerful way, suggesting a path toward progress, a notion that his opponents among the ancients would have vigorously rejected.[25]

The continuing censure of intellectual speculation from church officials added the zest of thwarting them to the zeal for discussing what human beings were now capable of explaining. The leading figures in the Enlightenment played their roles as liberators of thought with an admirable élan, moving ever closer to forbidden topics, such as the diverse ways of worshipping god.

Initially Amerindian religious practices had provoked horror, but after a century the association of them with the devil gave way to curiosity about exactly how and what the Indians did worship. During these same years Europeans were learning more about Eastern religions—Buddhism, Hinduism, and Confucianism. The Christian Church only tolerated discussion of diverse religious traditions if the unique access of the Church to speak for God was maintained.

Jesuit missionaries in Japan, responsible for much new information about Asians, had fought hard to make knowledge of other religions a springboard to recognition of the superiority of Christianity as the one true religion. The Jesuit Joseph-François Lafitau drew upon his deep knowledge of the customs of the Iroquois of North America to assert the commonality of a belief in a divinity and hence of the universal truth of core Christian values. On the other hand, critics of the clergy, if not of the church, were only too eager to draw new conclusions from the mounting evidence of spiritual diversity across the globe.

With increasing information about the varieties of spiritual regimens in the world, it was hard to keep at bay the thought that Christianity might be a particular local religion of Europeans—a truly heretical thought usually entertained by those who were already enemies of the church. Bernard Picart and Jean-Frédéric Bernard, two French expatriates enjoying the protected liberties of the Dutch Republic, used science to make this case. Their seven-volume *Religious Ceremonies and Customs of the World*, which appeared between 1723 and 1737, canvassed the vast accumulation of knowledge about religion for purposes of analysis and comparison.[26]

Unlike literary essays or travel accounts of eyewitnesses, *Ceremonies and Religious Customs* was footnoted like the learned tome that it was. Picart was a talented engraver. His illustrations of rituals, altars, and edifices were both sympathetic and authoritative. They formed an integral part of Bernard's text. Undertaken as a scholarly effort with the hope of commercial success, Picart and Bernard succeeded on both counts. The effect of seeing all religions treated sympathetically, with Christian practices examined along with the rest, had a subtle effect upon readers. While such a dispassionate approach might have outraged orthodox believers, others came to their own conclusions without any heavy-handed message from the authors.

Charles-Louis de Secondat, baron de Montesquieu, had taken on Christian beliefs more humorously in his epistolary novel, *The Persian Letters*, of 1721. Using the device of a correspondence between two Persians traveling through Europe, he succeeded in making Christianity seem the foreign faith. In this as in many other things, the Protestant break from the Catholic Church weakened Christianity's universal claims, contributing to speculation about the inherent diversity in divine practices. Montesquieu, whose wife was Protestant, even depicted kings, bishops, theologians, and priests as working the magic of wizards. David Hume, in his *Natural History of Religion*, made fear the driving force behind mankind's tendency to create deities. Relying on psychology more than the oppression of priests and monarchs to explain why human beings worshipped gods, Hume stressed that they were "oftener thrown on their knees by the melancholy than by the agreeable passions."[27]

These reassessments of Christian theology and cosmology stimulated new ideas about the origins of human societies. All the giants—Rousseau, Montesquieu, Adam Smith—joined in the task of integrating into the old accounts of human history what was now known about the world's peoples. A conjectural history emerged from this lengthy engagement with diverse human habits. The vertigo-inducing facts about marriage patterns, funereal customs, political systems, economies, and rituals of native inhabitants of the Western Hemisphere and Pacific Islands about which Gonzalo Fernández de Oviedo, Bartolomé de Las Casas, and Antonio Pigafetta had written, played a major role in prompting these ruminations.[28]

A cluster of Scottish thinkers—Smith, Hume, William Robertson, Adam Ferguson, Lord Kames—popularized the idea of successive stages of social development. They posited that human societies went from hunter-gatherers to a pastoral and nomadic stage to one marked by the introduction of agriculture, to be capped off with the final stage of commerce and manufacturing. Over the course of three centuries, travelers traipsing around the globe had encountered people existing in all four of these stages. More sociological than historical, this schema wrote progress into human destiny for all peoples, even those still hunting and gathering for their food. Conjectural histories convinced readers because they presented a plausible, universal explanation of obvious changes. "The Great Map of Mankind is unrolld at once," Edmund Burke exclaimed.[29]

A part of the new, modern consciousness was a heightened awareness that society had not been handed down by some divinity, but was rather the product of human beings themselves. The word "society" itself had gained currency as a reference to the presence of informal social relations. Urban sociability and the market economy had created this new space during the course of the eighteenth century. Both voluntary and uncoerced, it was lodged between the philosophers' state of nature where, as Thomas Hobbes famously remarked, life is "solitary, poor, nasty, brutish, and short" and the rule-regulated regime of a formal political order.

Adam Smith in his great work *The Wealth of Nations* made the division of labor and the extension of the market the driving forces behind the movement to the final stage of commerce. (He apparently picked up the famous description of pin making in Diderot and d'Alembert's *Encyclopédie*.) Commercial society, according to Smith, depended upon the prior accumulation of capital because installing the division of labor required substantial investments in labor.[30] Aligning economic development with nature, all of the Scotsmen wrote history with policies of free trade in mind. But it's well to remember that economics at that time had no separate identity as a study. Smith held the University of Glasgow's chair in moral philosophy, which then included natural theology, ethics, jurisprudence, and political economy.

Sophisticated inquiries abounded, and experts devised ingenious ways to resolve their disputes about such things as the shape of the earth. These successes contributed powerfully to the idea of sustained improvement. The concept of successive stages helped maintain the oneness of humankind while recognizing its various expressions around the globe. History could now be seen as a process. This view is familiar to us, but represents a decisive break with the narratives of the Bible and accounts of dynasties and wars.

The idea of placing economics in a natural perspective startled contemporaries who were used to considering economics the handmaiden of politics, necessarily so with the international rivalries of the time. Smith and others had also dispensed with Providence, a venerable way of searching events for messages from God. Not knowing exactly how human groups had changed, the Scots constructed hypothetical accounts. Traditionalists were stuck with the Bible, but it no longer convinced everyone as a guide to actual events. The new conjectural history depicted change through universal stages rather than presenting a particular narrative like that in Genesis. Once again the imposition of order on the proliferation of information about various societies turned out to be a challenge to religious orthodoxy. In the seventeenth century Bacon had considered belief in historical cycles an obstacle to the flourishing of scientific inquiry.[31] Relief had come at last. Inquiries prompted by the discoveries made by explorers and naturalists had led to a new conception of history and the social potential of human beings.

In the eighteenth century European governments awakened to how further explorations would enhance their power. Spain, France, and England set up handsome prizes for someone who could devise a way to measure longitude at sea and reduce the losses they all incurred from faulty navigation. The French government funded the expeditions to the equator and Arctic Circle to disprove an Englishman's theory about the shape of the earth and incidentally improve navigation. The British Navy sent exploratory voyages into the Pacific. Not to be outdone, Napoleon arranged for a cadre of 160 scholars and scientists to follow him when he invaded Egypt in 1799. The curiosity that had stirred explorers, travelers, naturalists, and their readers now found an ally in chauvinism as national rivalries replaced the religious enmities that once dominated European affairs. Curiosity, redefined as scientific inquiry, had acquired powerful patrons.

CHAPTER SEVEN

ATTENTION TURNS
TO THE PACIFIC

*Engraving by an unknown artist depicting Joseph Banks showing
a group of Tahitians the planet Venus on the Sun, during Captain James Cook's
first voyage to the Pacific to observe the transit of Venus, 1769.*

Although explorers had circumnavigated the globe two dozen times by the end of the eighteenth century, the Pacific was still virtually unexplored. Landings had been made on New Zealand. The Philippines had been colonized, but Europeans had yet to visit the Arctic Circle and most of the islands that the Polynesians and Melanesians occupied. (For that matter, the interiors of Africa, South America, and North America west of the Mississippi River had hardly been touched by Europeans.)

But out of sight did not mean out of mind. Spain, France, and England—even Russia—wanted to map the lands in this vast body of water. Once Magellan reached the East Indies by sailing west, speculation about a possible continent in the southwest corner of the ocean became intense. Actually, Marco Polo had started the interest in Australia with his tale of a place southeast of Java, distinguished by a surfeit of gold and hostile inhabitants. He would be right about both, though it wasn't until an Australian, returning from California in 1850 with the determination to find gold in his own country, discovered it within a week.[1] He had learned that the presence of quartzite was an indicator of goldfields.

Late sixteenth-century cartographers obligingly located this golden spot, which they called Beach, along a 2,000-mile arc extending south of Java to south of the Solomon Islands in the Coral Sea. The Great Southern Continent or *terra australis incognita* became a magnet for the imaginative. If there were indeed a continent, what riches might be found there or even introduced, as sugar had been to the Caribbean? The absence of full information didn't stay the hand of mapmakers, who began charting every sighting of a landmass that a European Pacific voyager reported.

Imperial ambitions combined with scientific curiosity in matters concerning the Pacific. National treasuries were opened to pay for expeditions, whether organized by their own navies or a royal academy. Voyages proved much more efficient than land expeditions, which were more arduous and dangerous. Ships, with their cabins for astronomers, botanists, and draftsmen, not to mention cargo space for specimens and drawings, proved to be wonderful floating labs.[2]

Twenty-six European explorers circumnavigated the globe between the voyage of Magellan's surviving ship in 1519 and that of Fedorovich Kruzenshtern and Yuri Fedorovich Lisianski, who sailed for Russia in 1803. Reviewing them gives perspective to Pacific exploration. Because they couldn't establish longitude accurately in the sixteenth century, neither the Spanish nor the Portuguese knew exactly which part of the Pacific the pope had given them. This didn't keep Spain from claiming sovereignty over the whole Pacific, which is why the king sent out a second fleet after Magellan's. This second, little-known effort was even more plagued than the first. Nine years of wrecks and wrangles with the Portuguese winnowed the 450 crew members to eight. None of the seven ships survived.

Evidently unimpressed by the pope's line of demarcation dividing the world between the Iberian powers, England, Holland, Italy, France, and Russia sponsored the next twenty-three voyages with the lion's share of twelve coming from England. Francis Drake completed the third circumnavigation over a half century later. There were two French circumnavigations and one each from Italy and Russia. The solo Italian venture had the distinction of carrying the first commercial, round-the-world passengers.[3] Thomas Cavendish was the second Englishman to circumnavigate the globe; he arrived back in England just after the Spanish Armada had been dispatched in the fall of 1588. He sailed up the Thames in great splendor with gold and blue silk sails. Even more spectacularly than Drake, he attacked the Spanish and their monopoly pretentions, bragging to his patron, Queen Elizabeth, that he had looted, burned, and sacked his way across the Pacific in defiance of the Spanish. He entertained the queen on board his ship, the *Desire*, and she knighted him. A century and a half later the English would return to the area in earnest. Word of these voyages spread across Europe through the thirteen travel journals that were subsequently published.

More to the point of imperial designs, twenty years earlier Philip II of Spain wanted to occupy the impressive archipelago where Magellan had lost his life, so in 1564 he had commissioned an expedition to do this. Successful in subduing the native inhabitants, he named the islands the Philippines after himself. Philip's goal was to establish trade between Asia and Acapulco on the west coast of Mexico with the Philippines as its hub. A western return route now became imperative; it was far easier to go from east to west in the Pacific, as Magellan had done. Philip picked Andrés de Urdaneta to find a route back, and he wasn't disappointed. An ingenious navigator, Urdaneta speculated that perhaps the tides in the Pacific were like those in the Atlantic, with winds blowing west at the northern latitudes. He sailed to 38 degrees north and caught westerlies that carried him to the coast of North America, near Cape Mendocino in California, in 1565.

The Spanish settled into exploiting what quickly became the most important center of Asian trade. Spices, silks, ivory, porcelain, and textiles from China, Japan, and the Moluccas came to the Philippines and then, using Urdaneta's route, were transshipped to Mexico and from there to Peru and Spain. Loaded with silver, enormous galleons sailed annually from Acapulco until the Mexican Revolution brought the 256-year-old trade to an end. There was a rationale behind the Manila galleons' size. The merchants who ran the royal monopoly out of Seville secured a decree limiting the traffic to two ships sailing each year from either port. Even with a protecting armada, these grand galleons fell prey to the English and Dutch who tapped into Spain's riches by raiding the fleet, sometimes in cooperation with one another.

Still, Spain continued to explore the Pacific, always in hopes of finding that continent laden with gold that Polo had mentioned. Pedro Fernandez de Quiros, a Portuguese sailing for Spain, landed in the New Hebrides in 1606. Luis Vaez de Torres, who had sailed first with Quiros, returned to the area in search of Australia. He discovered New Guinea and the strait between New Zealand and Australia that now bears his name, but he left no record of sighting the prized subcontinent. That accomplishment went the same year to a Dutchman, Willem Janszoon, who made landfall in Queensland.

Still the geography of Australia remained uncertain. On a second voyage, Janszoon failed to circumnavigate it and departed believing that he had found a large island. Sailing with the Dutch East India Company, Dirk Hartog made the next landfall in Australia in 1616, this time at various sites on the western coastline. Twenty-six years later, Abel Tasman was dispatched to settle the question about Australia's dimensions. On his way he sighted a new island, which eventually bore the name of Tasmania. Thrown off course after leaving there, he spotted New Zealand, which Tasman assumed to be an island near the southern tip of Argentina. No one yet had grasped the vastness of the Pacific.

Sailing back to Batavia (today Jakarta), the headquarters of the Dutch East Indies, Tasman passed the Fiji Islands. Despite these new discoveries, both of Tasman's exploratory voyages were accounted failures because neither gold nor favorable new trade was forthcoming. Europeans ignored the area for almost another century. Australia, generally referred to as New Holland, became a fading memory until the early eighteenth century, when a sixty-two-year-old Dutch lawyer sailing for the Dutch West India Company took a fleet of three ships down the coast of South America with the intent of finally finding it.

Jacob Roggeveen was actually fulfilling the ambition of his

father, who had been a distinguished mathematician with keen interests in astronomy, geography, and the theory of navigation. Roggeveen didn't come close to Australia, but he did come upon an inhabited island in the Pacific on Easter Sunday, 1722. Naming it Easter Island, he sailed on to Batavia, discovering Samoa on his return voyage. Sixteen years later, the French East India Company sent a fleet of two ships to find a trading base for their voyages to the East Indies. Jean-Baptiste Bouvet de Lozier led what became another miserable failure, though he did earn his spurs as a discoverer when he found an uninhabited volcanic island in the South Atlantic that bears his name today. Bouvet reported back that the "Terres Australes" were much farther from the South Pole than imagined.

Meanwhile Russians stationed in Siberia began searching for rivers that led into the Arctic Circle. Semyon Dezhnev reached the easternmost tip of Asia, but he was neither a skilled navigator nor a geographer. What he wanted and found were sable fur, walrus ivory, and silver ore. After a hazardous excursion to the tundra in seven small rowboats, Dezhnev and his few surviving companions found a land route that proved to be the superior way to the Anadyr River, which flows into the Bering Strait. It became the main route from the Arctic to the Pacific and Kamchatka, the large peninsula that faces Alaska. Dezhnev did succeed in reaping the profits from two tons of ivory after coming upon a walrus rookery. Whether or not he actually passed through the strait that separates Asia from North America, which Vitus Bering found eighty years later, is in some doubt. Bering's 1742 expedition made Russians aware of the rich trove of seal and otter skins to be found in Alaska, prompting them to establish permanent trading posts there. In 1867, when the trade in pelts slumped, they sold Alaska to the Americans, a purchase known as "Seward's Folly" after Abraham Lincoln's secretary of state, William Seward, who negotiated the treaty.

The rivalries among European powers led to five successive wars during thirty-eight years of the eighteenth century. As befits nations with global reach, France and England confronted each other in five different places on the globe, from India to Hudson Bay. Privateers, licensed during wartime by their governments, often turned into pirates when peace treaties were signed. By the beginning of the eighteenth century, merchants had had enough of this nonsense. They used their influence in Parliament to get the British Royal Navy to free the seas of marauders. Captain Kidd became the poster boy for the new policy when he was captured in the Indies and brought back to England for a showcase trial and public execution in 1701. European countries were willing to fund exploratory expeditions, but they could no longer tolerate interference with the global trade that sustained employment at home and brought in new revenue. The buccaneer years were drawing to an end.

No tale from the Pacific stirred as much popular interest as that of Alexander Selkirk, first mate in a privateer searching for Spanish booty in 1703. After his ship landed on an island off the Pacific coast of South America, Selkirk became convinced that the boat was unseaworthy. He begged the captain to abandon the ship and wait for another to pick up the crew. Instead the captain abandoned him. Alone on the island, Selkirk fended for himself for four years, taming cats, eating goats, evading the Spanish, reading his Bible, and searching the horizon for sight of a friendly ship. He was rescued four years later by another English privateer. Sound familiar?

This is the real-life story that probably inspired Daniel Defoe to compose *The Life and Strange Surprising Adventures of Robinson Crusoe*. Needing money to pay for his daughter's wedding, Defoe created a story so compelling that it has never been out of print since. He gave his marooned mariner an escaped prisoner, whom Crusoe named Friday, to ease his loneliness, and invented a lively cast of cannibals, captives, and mutineers to enliven the plot. Selkirk, by the way, had been correct. The ship he left sank a month after sailing away.

Curiosity had obviously acquired a lot of new partners: vying nations, avid merchants, ambitious naval commanders, and expert naturalists homing in on particular problems. Strategic and commercial motives reinforced one another. When the French and English weren't at war, they were sending out expeditions to the Pacific to lay claim to its unknown treasures. Two of the most famous eighteenth-century navigator-explorers, Louis Antoine de Bougainville and James Cook, acquired much of their sailing skill as navigators fighting on opposite sides in the Seven Years' War. At first rival explorers, Cook and Bougainville later competed in a race to get their journals published and translated into each other's language. A curious public had created profits and acclaim not just for the adventures but for the stories about them.

As something of a gesture to recover the national pride dented by defeat in the Seven Years' War, the French government sent Bougainville around the world in what was one of the most memorable voyages of all time. Bougainville had already distinguished himself as a mathematician with a publication on integral calculus. This earned him membership in the English Royal Society while he was living in London as secretary to the French embassy. His subsequent circumnavigation of the globe was successful on several fronts. It was France's first, and he lost only seven of his two-hundred-plus seamen and neither ship. Two and a half centuries earlier Magellan's 270 men on five ships dwindled to eighteen pitiful survivors on one leaky vessel, not to mention the failed second Spanish effort.

More important than making the fourteenth circumnavigation of the globe was Bougainville's stay in Tahiti. He claimed it for the French in 1767, though unbeknownst to him, two English explorers had preceded him there. Halfway between South America and Australia, Tahiti stretches 120 miles just below the 17th parallel. Bougainville introduced a new Garden of Eden to eager European readers, a place where plenty had banished labor and tropical beauty reached from humans to plants to black volcanic beaches to arresting mountains. Bougainville's ship *Boudeuse* carried an astronomer aboard along with a well-known botanist, Philibert Commerson. None of them were prudish, so they responded to the Tahitians' easy, sensual living with appreciative delight.

Commerson's valet on the trip was Jeanne Baré, who dressed in male clothes. When Baré went ashore, the Tahitians gathered around, shouting that she was a woman. How they immediately discerned her sex, which had eluded her French compatriots living in close quarters for almost five months, remains a mystery. Baré told Bougainville that the prospect of going around the world had "raised her curiosity." Bougainville dryly commented that once her gender was disclosed, "it was difficult to prevent the sailors from alarming her modesty," but Baré completed the trip and became the first female circumnavigator.

Commerson found many new plant specimens on the trip, including two that would spread to the New World. He rewarded the *Boudeuse*'s captain by giving his name to the gorgeous bougainvillea shrub with blossoms in luscious colors of lavender, coral, magenta, and cream. Commerson also introduced Europeans to the hydrangea, which he found in China.

Bougainville supplied a new specificity to the old trope of the noble savage. Some scholars trace the concept back to sixteenth-century Spanish chroniclers like Las Casas, who claimed that God had created the indigenous people of the Caribbean "without evil and without guile," but that image died with better acquaintance of Amerindians, especially the Brazilian cannibals. Inspired by Tahiti, Jean-Jacques Rousseau and Denis Diderot used native people to hold up a critical mirror to their own overly refined societies. Bougainville's journal, *Voyage autour du monde*, published in 1771 with an English translation in 1772, started what might be called a rage over Tahiti. Its descriptions of a beautiful and generous people living in leisure on a stunning island of luxuriant valleys and majestic mountains enthralled readers. The salubrious climate stirred envy; the flowering plants, delight; the people's hospitality, indulgent respect.

Subsequently Bougainville learned that Tahiti was not quite the paradise that he at first imagined, but his journal's depictions of an erotic paradise with its enticing pairing of native beauty and free love endured. Reports about Tahiti came at the dawn of the romantic movement in art and literature, making it almost irresistible not to see the mountain-crested island as an unspoiled Arcadia. Indeed, Bougainville named it La Nouvelle Cythère, the new island of love, though the more prosaic English had called it King George III's Island.[4] Successive accounts poured in about Europeans taking in the pleasures of the Pacific. Ribald commentary quickly ensued. Tahitian morality began to attract the attention of missionaries, who followed in the wake of the first voyagers. Then the innocent joy that men and women expressed was reconceived as a kind of corruption stemming from the easy conditions in their Edenlike seclusion.

But there was more to be found in the Pacific than tropical islands. Dalrymple, the admiralty geographer, was translating some captured Spanish documents when he came upon Torres's old testimony about the strait south of New Guinea. Dalrymple had become convinced that an Australian continent existed and that England should possess it. His supporting evidence included two diaries, published in 1703 and 1709, from William Dampier's voyage to Australia. Dampier combined a passion for botany and zoology with a career as a privateer. He did the first serious investigation of the flora and fauna of Western Australia, bringing back a wide range of specimens and extensive notes on coastlines, trade winds, and sea currents for the area around Australia and

New Guinea, but he hadn't gauged its size by encircling it. Dampier was outstanding as both the first three-time circumnavigator of the globe and an adventurous naturalist. His popular travel journals introduced the words "barbecue," "chopsticks," and "avocado" to the English public. Some consider him the inspiration for Robinson Crusoe; others for Samuel Taylor Coleridge's poem, "The Rime of the Ancient Mariner."

In Dalrymple's hands Dampier's journals became fodder for a campaign to get the British government to locate and claim the great subcontinent in the South Pacific. With these and other documents that he had assiduously assembled, he published his *Historical Collection of the Several Voyages and Discoveries in the South Pacific Ocean.* Dalrymple's obsession finally aroused the British Navy, but he failed to gain command of the exploratory expedition. The admiralty would only put one of its own in charge and so chose James Cook, a relatively junior naval officer who emerged from obscurity to become a national hero as the quintessential explorer of the Pacific. For most people Cook's first name is "Captain," a rank he earned only after his second voyage to the Pacific.

Cook was an ambitious autodidact. With just five years of primary schooling, he taught himself geometry, algebra, trigonometry, and astronomy before joining the navy at the age of twenty-six. He quickly became skilled in marine surveying and distinguished himself as a navigator in the Seven Years' War. More important to his future, he executed a masterly mapping of the ragged coastline of Newfoundland. Cook was a man of great physical courage and personal discipline. These qualities didn't go unnoticed. Between 1768 and 1776 he led three extensive voyages to the Pacific paid for by the British Admiralty and the Royal Society in a partnership of government and science. Events would reveal Cook's indomitable spirit as well.

Occasionally, curiosity homes in on a particular question. In the eighteenth century one of them was the size of the solar system. Johannes Kepler's study of planetary orbits in the seventeenth century had given eighteenth-century astronomers the ratios of the distance between the planets. To calculate the distances among them, they needed a celestial indicator. Kepler had predicted the dates when Venus would pass directly between earth and the sun. Fifty years later, the British astronomer Edmund Halley figured that measuring the transit of Venus across the face of the sun could establish the distance between it and the earth. From this figure could be calculated the relative distances of the other planets and hence the size of the solar system as a whole.[5]

A French astronomer, Jean-Baptiste Chappe, had succeeded in making excellent observations of the 1761 transit of Venus from central Siberia. The 1769 transit held out even greater promise. Venus's eclipse of the sun occurs twice every century and wouldn't happen again until 1874 (Halley had died in 1742). Because it was important to observe it from widely different parts of the world, the British and French academies drew on astronomers from Russia, Germany, Scandinavia, Italy, the Netherlands, and the American colonies to take the very finest equipment to carefully chosen spots to capture Venus's transit across the face of the sun.

Despite this remarkable international endeavor, few of the calculations from the Philippines to central Russia to Hudson's Bay jibed, although America's Benjamin Rittenhouse was able to calculate the distance between Venus and the earth at 95 million miles, only 2 million miles off. Chappe himself died on his way home from Baja California.

Cook's auspicious career as the Pacific's greatest explorer began with this astronomical event, which gave the British Navy its ostensible motive for sending the *Endeavour*, with Cook in command, to Tahiti, which Samuel Wallis had discovered two years earlier. Made even more famous by Bougainville's visit, Tahiti became the Royal Navy's observation spot. After an uneventful voyage, Cook delivered Charles Green, an astronomer at the Greenwich Royal Observatory, to the island. The navy also wanted to test a chronometer modeled after Harrison's prizewinning clock. Cook named the archipelago, which included Tahiti, the Society Islands in honor of the Royal Society, which had done so much to keep British officials committed to Pacific exploration. The *Endeavour* also carried a contingent of naturalists and artists whom the seamen called the "experimental gentlemen."[6]

"Experimental gentleman" hardly does justice to Joseph Banks, the twenty-six-year-old star in London society's firmament who led the naturalists. Musing that there could be a better motive for so adventurous a voyage than tracking the transit of Venus, Bank had proposed that the Royal Society sponsor "a Voyage, as a Voyage of Mere Curiosity," adding that "such a voyage seems at least as interesting to Science in general and the increase of knowledge as the Observation which gave rise to the Present one."[7] Well-connected and exuberantly social, Banks had been obsessed with botany since he happened upon a field of wild flowers at age fourteen. Full of himself and his hopes for science, Banks put together a team of botanists and artists to collect specimens that would create a powerful graphic record of the people and places they visited.

The English post on Tahiti acquired the name of Fort Venus. The starchy Captain Cook attributed the name to the voyage's mission; his junior officers were more inclined to attribute it to the goddess of love so conspicuous in Tahiti. In a three-month stay, the crew discovered all the delights that Wallis and Bougainville had already advertised. For their part, the Tahitians, devoid of minerals, found it irresistible to steal English metal in tools, weapons, and equipment. The two-ship fleet had not been there long before the price for making love to a Tahitian woman had been set at one nail, an arrangement that led an enterprising seaman to steal a hundredweight bag of them. Cook barely tolerated the sexual frivolity, but he was adamant about protecting the ship's metal fittings, recalling that Wallis had almost lost his ship in a storm on the way home, so bereft of nails were its timbers.[8]

Banks's diligence in scouring Tahiti for botanical specimens prevented him from "going native," but he spent more time than anyone else in their party living among the Tahitians, whose generosity he both celebrated and reciprocated. Before departing at

Sydney Parkinson's portrait of a Maori warrior drawn when Captain James Cook's first Pacific voyage reached New Zealand, 1770.

the end of the expedition's three-month stint, Banks distributed
to the Tahitians seeds for lemons, limes, oranges, and watermelons and took with him a fine array of native flora and fauna. He
also took back to England a wise priest who had befriended him,
along with the priest's son. Banks's *On the Manners and Customs of the*
South Seas, composed from his journal after his return, added some
substance to the fantasies about Tahiti with sympathetic accounts
of the Tahitians' fishing, cooking, drum making, boatbuilding, and
dancing. He had got to know them so intimately that he was even
able to inquire about the infanticide they practiced to restrain
population growth, the only thing that truly shocked him.[9]

While Cook's voyage was ostensibly to sail to Tahiti to observe
the transit of Venus, navy officials also wanted to get a bead on
the mysterious southern landmass that had so long intrigued
Europeans. They instructed him to do astronomical, hydrograph-
ical, and geographical work, as well as make inquiries into soils,
vegetation, fauna, and marine life. Upon leaving Tahiti, Cook
opened his secret orders and learned that he was to search for
the continent in which Dalrymple had such faith. The many
Dutch landings had all been on the west coast, leaving up in the
air exactly what the dimensions of this sizeable area might be. A
friend of Dalrymple, Banks thrilled to the idea of finding Australia. Cook did not believe in its existence, but followed his orders
and plunged south.

While Dalrymple's ambition to lead the Royal Navy's expedition was frustrated, something of his did make it on board the
Endeavour. An account of the discoveries made in the South Pacific Ocean was
a magnificent collection of the maps with the routes taken by
every known explorer in the region. With these Cook plied what
seemed like waters devoid of life for three months before reaching New Zealand and learning just how hostile the Maoris could
be. He displayed the sangfroid of a sophisticated man when he
discovered their cannibalistic ways. In their negotiations, Banks's
Tahitian priest proved an invaluable translator. For Cook the

challenge was to survey the coastline of New Zealand while circumnavigating it to prove that it was an island and not part of
a continent.[10]

After mapping New Zealand through gale-force winds and
plunging temperature, Cook steered the *Endeavour* westward and
discovered to his amazement the long-sought east coast of Australia, which he claimed for Great Britain. It took another four
months to do the mapping, during which the English were introduced to the novelties of aborigines who painted their naked
bodies with white stripes and animals that moved by making
enormous jumps with their hind legs. A year after the secret
instructions were opened, Dalrymple's maps and Cook's superior
navigating had led them to Australia. Not until 1803 did Mat-
thew Flinders triumph over the treacherous reefs in the area to
complete a circumnavigation of the subcontinent, fifteen years
after the English first settled the hospitable Botany Bay in New
South Wales, which Cook discovered and named.

Daniel Solander, one of Linnaeus's students, assisted Banks's
work along with the Finnish botanist Herman Sporing. Solander
did a great deal of plant collecting, particularly when the *Endeavour*
was in Australia, but Banks got credit for introducing eucalyp-
tus, acacia, and mimosa to Europe. Banks took very seriously the
advice of previous scientific travelers to bring along trained draftsmen, and, if possible, artists. He took both. There were three:
Alexander Buchan for dilettantes, Sydney Parkinson to provide
sketches of new species, and the botanical draftsman Sporing to
capture the landscape and the exotic people in it. The full retinue consisted of eight persons and two dogs. Parkinson became
very skilled at producing topographical accuracy, encouraged by
Cook's persistent demand for accuracy in all things. Banks was
the energetic genius behind the voyage's many-pronged initiatives
in natural history. As he noted, he was the first man with a scientific education to complete a voyage of discovery, a feat that the

great French naturalist, Georges Cuvier, anointed as "an epoch in the history of science."[11]

Cook proved to be an unusual commander in more than technical ways. He took an unprecedented interest in the health of his crew. His experience in Newfoundland convinced him that a proper regimen could keep his seaman alive. He took along sauerkraut as a specific against scurvy, making his officers set an example by eating it. When the sauerkraut ran out, he gathered celery and grass as substitutes. He insisted on cleanliness and exercise as well. On the voyage between England and Tahiti, only five men perished, an impressive record for the time. Forced to stop for repairs in pestilential Batavia on the way home, Cook's men got trapped in that notorious den of death, sauerkraut having little effect on dysentery and malaria.[12]

Sporing and Parkinson were among the more than thirty men lost to disease in Batavia, along with the two Polynesian passengers, so gloom hung over the *Endeavour* when it docked at the Downs in England in July 1771. Within a few weeks, the public got wind of the arrival of the "experimental gentlemen" and began celebrating them as heroes. King George III received Cook, Banks, and Solander at court. For Banks this was the beginning of a friendship with the then young king who shared his love of horticulture. Cook returned to a quiet reunion with his family, while Banks set London ablaze with excitement over his dashing scientific adventure. His days were crowded with state dinners, appearances at court, and public lectures, not to mention an honorary degree from Oxford, which he had left, degreeless, seven years earlier. The admiralty was thrilled with the forty islands that Cook had discovered and the geographic speculations his voyage laid to rest, but it was Banks who monopolized the glory.[13] Lionized wherever he went, he achieved rare fame for someone not yet thirty.

The admiralty commandeered all the journals that had been

kept on the voyage. It was eager to get Cook's in print to secure British claims to his discoveries and to take some of the public attention away from the English translation of Bougainville's journal, which had just appeared. The admiralty hired John Hawkesworth, a well-known journalist, to give Cook's writing the literary polish it lacked. Hawkesworth also had Banks's journal, which delighted the editor because of the lively, even juicy, details that the account of the chaste Cook lacked. Hawkesworth, by turns morally incensed and annoyingly prurient, produced a doctored mélange of records. It humiliated the punctilious Cook, so he secured a promise from navy officials that he would have control over any future journal publications. James Boswell, the great biographer of Samuel Johnson, told Cook that Hawkesworth had used his journal as "a London Tavern-keeper does wine," by brewing it. Hawkesworth's depiction of the lascivious sexual exploits of Banks became something of a trial for Banks when London wits derided him as a "botanical libertine." His fascination with botany was insidiously linked to an overactive libido.[14] The Linnaean tradition persisted.

Publication of the translations of Bougainville and Cook's journals ramped up the British-French competition while intensifying a general interest in the new site of European explorations. Louis XVI had been so enthralled reading Cook's *Voyages* that he sent out a French expedition to explore the Pacific with explicit instructions to investigate "the character, manners, customs, bodily constitution, language, government and number of the inhabitants" while collecting "garments, arms, ornaments, utensils, tools, musical instruments, and everything used by the different people he visited."[15] Headed by Jean-François de Galaup, comte de Lapérouse, the expedition of 220 men set sail in two ships in 1785. More comprehensive than other expeditious, the proposed circumnavigation of the globe began with stops in Easter Island, Hawaii (the first European visit to Maui), then sailed to Alaska, down the coast of California, and across the Pacific Ocean to Macau, where the French sold the furs they had collected in Alaska. From there Lapérouse took his ship to the Kamchatka peninsula, where he received instructions to visit the new British settlement in Australia.

Nothing further was heard from any member of the Lapérouse expedition after its stop at the British colony at New South Wales in January 1788. A rescue was set in motion in 1791, although France was in the middle of a revolution. Not until thirty-five years later did an Irish sea captain sailing in the Solomon Islands find remains of the ships between some coral reefs. Subsequent investigative voyages involving dozens of scientists in 1964, 2005, and 2008 were able to identify Lapérouse's ships. The ill-fated voyage also contains a fascinating "what might have been." A sixteen-year-old Corsican cadet named Napoleon Bonaparte applied to join the trip, but failed to make the last cut when the crew was chosen![16]

Although Cook's voyage had demonstrated that Australia was as large as a continent and not attached to New Zealand, there were still those—particularly the persistent Dalrymple—who believed that the long-sought *terra australis incognita* lay even further south. The only way to still these doubts would be to send Cook out on a second voyage. Banks very much wanted to be a part of the next expedition and began planning how best to enlarge the assigned ship, the *Resolution*, to accommodate the many artists and scientists—and two musicians—he planned to take along. Never one to exercise restraint, Banks's successive demands exhausted the patience of the admiralty, which finally balked. His patron, the Earl of Sandwich, sided with the navy, and Banks withdrew in a huff.[17]

At twenty-nine Banks was too young to be discouraged by one setback. With a large fortune at his disposal, he went off to Iceland in a private expedition. He sent Polish botanist Anton Hove to India to collect medicinal drugs and cottonseed as well as to observe Indian methods of cultivating and manufacturing cotton textiles. He placed botanists on voyages scouring for plants in the Arctic, Brazil, Australasia, the West Indies, and Central America. He had tea plants and hemp shipped from China and, with a keen sense of commercial possibilities, he got hold of the Central American host plant for cochineal, the insect from which crimson dye is extracted. Banks believed that species from one tropical location could grow in another, a proposition that was soon to be tested with initially disastrous consequences.[18]

Acting as an informal adviser to George III for the Royal Botanic Gardens at Kew, Banks was able to go even farther afield for botanical specimens. All redounded to his fame and that of the Kew Gardens. Just seven years after his return with Cook, the Royal Society elected Banks president, a position he held for forty-one years. Carl Linnaeus offered him the entire collection of specimens, books, and manuscripts that his father and grandfather had amassed during their careers as distinguished botanists. When Banks turned it down—there's just so much one man can own—one of his friends, James Edward Smith, snapped it up for the bargain price of a thousand pounds sterling. Himself a botanist, Smith founded the Linnean Society, which has played a supporting role for naturalists ever since. (The Linnean Society spells its name differently from other references to Linnaean taxonomy.)

With Banks and his retinue eliminated from his plans, Cook could get the *Resolution* back in shape for a voyage to resolve the puzzle about possible continents in the South Pacific. The admiralty added a second ship, the *Adventure*. Each boat had its own astronomer to observe longitudes and latitudes and test the accuracy of the chronometers they took along. In Banks's place were two German naturalists, Johann Reinhold Forster and his son, George, who later took Alexander Humboldt, one of the greatest nineteenth-century explorers, under his wing. At the Cape of Good Hope, Cook added Anders Sparrman, another of Linnaeus's students, to his scientific complement. Thirteen-year-old midshipman George Vancouver, slated for a grand career as the explorer of the northwest Pacific, was also part of the crew.

Harrison's chronometer, for which George III had secured the longitude prize, served Cook well; he could now calculate his longitude as he sailed into unknown waters.[19] Most of the European crossings of the Pacific had followed routes close to 40 degrees south. Cook decided that sailing farther south would clinch the argument about the location of a second southern landmass, so he plunged south from Cape Horn almost 1,500 miles to 60 degrees south. Before leaving the South Atlantic, he claimed South Georgia Island for Britain. Circling the globe in the most southerly latitude yet, Cook expressed a desire not just to go "farther than any man has been before me, but as far as I think it is possible for a man to go."[20] Cook took the *Resolution* into the Antarctic Circle repeatedly, with the *Adventure* following, both ships defying the frigid temperatures and fierce winds in pursuit of truth about the southernmost waters of the Pacific.

Banks's obsession with paintings to please as well as record was not forgotten on the second voyage. An established British artist, William Hodges, went along. The spectacular views on offer during the voyage inspired Hodges to lift the sedate landscapes of eighteenth century classicism to a new level of excitement. He painted large-scale canvases with arresting views of the Pacific and Antarctica that emphasized Cook as a discoverer. Stirred by his experiences at sea, Hodges influenced the sailors who were assigned to draft oceanic views to experiment with lighting and extend their range of media from pencil to ink, and to water- and oil colors. Occasionally these heightened perceptions of light enabled the artists aboard to discern the presence of treacherous shoals. They also portrayed the artifacts they encountered. The whistles, fishhooks, canoes, tattoos, headdresses, and statuary of Pacific native art often became the focus of the Europeans' paintings.[21]

Cook got back to England in the summer of 1775, after almost three years away. This time the accolades were all for him. With a different component of experts and artists he had circled the globe at the most southerly latitude yet. While he didn't touch Australia again, he had proved conclusively that no landmass existed in the southern waters. It was a masterful achievement. The admiralty promoted Cook to the rank of captain, and George III spent an hour conversing with the person who had gone "as far as possible for a man to go."[22]

Cook was handsomely pensioned off, but within six months he jumped at the chance to lead one more expedition. The goal this time was to find a passage across the northern reaches of North America. Conviction that such a passage existed rested on the erroneous idea that salt water did not freeze. Hence the ice caps at the North Pole must have come from freshwater rivers, which could provide a passage with proper navigation. A northwest passage would be of so much strategic value to Britain that a handsome prize of twenty thousand pounds had been established for the lucky discoverer. While Americans were declaring their independence, in July 1776, Cook set sail to search for the northwest passage. It would be his last voyage.

The third voyage was very much a navy affair without the trappings of science and art, save for one illustrator, a young botanist from Kew Gardens, and one astronomer, about to be rendered useless by the proven accuracy of Harrison's chronometer. Twenty-one-year-old William Bligh, a gifted surveyor who had been at sea since he was seven, joined the crew as sailing master. With two ships, the *Resolution* and the *Discovery*, Cook headed out from Europe to the Pacific. George III had emptied his menagerie so that Cook could carry goats, sheep, rabbits, chickens, pigs, cows, bulls, even two peacocks, for the crew's delectation.

After touching down at Tahiti, Cook sailed north through what he expected to be open sea. Instead he came upon a beautiful group of fertile volcanic islands stretching from the northwest to the southeast across the Tropic of Cancer. They teemed with people and vegetation and glorious landscapes like those in the Society Islands. When he came ashore, the Hawaiians greeted him like a god, which apparently they had decided he was. This designation didn't slake their interest in the tools and instruments the English had. Like the Tahitians, they appropriated them whenever possible. To the amazement of the English, the Hawaiians understood the Polynesian tongue that some of the seamen could speak, just as had the people of New Hebrides. These sea-migrating people, thousands of years earlier, had ranged widely in the Pacific after they left Asia.

Despite the allure of Hawaii, Cook didn't tarry. After a fortnight spent resupplying with baskets of yams and fish and barrels of water from the plentiful freshwater streams, he sailed north to scour the heavily indented northwest coast of North America for any signs of a passage. For eight months Cook mapped the area from Nootka Sound to the Bering Strait, dipping into hundreds of coves and inlets, but to no avail. The crew felt the disappointment of failure as keenly as their captain, for they would share in the prize were they to find a water passage leading to Hudson's Bay. They were also discontented, having traded the balmy air of Hawaii for the chilly winds of North America, indulgent recreation for arduous effort. Cook himself was testy with his men and those he encountered. His sure command, which had been remarkable before, dissolved into an irritable insistence upon obedience. He celebrated his fiftieth birthday correcting old Russian maps and no doubt cursing them for luring the British thither. Finally giving up when a consensus of his officers echoed his conviction that there was no passage, they headed back to Hawaii and disaster.

Leaving the area as winter's frosty curtain descended, Cook

went back to the tropical archipelago he had discovered almost a year before, this time to the big island of Hawaii. Hundreds of men and women rowed out in canoes laden with food and flowers to greet the two British ships. When Cook went ashore, again the people prostrated themselves before him. They led him up to a sacred stone for a ceremonious welcome. The Hawaiians treated him like a superhuman being, and Cook didn't resist, as the captain of the *Discovery*, Charles Clerke, did.

Undeflected from his goal, Cook planned one more attempt to locate a northwest passage. Shortly after departing, the *Resolution* sprung its foremast, requiring a return—untimely, as it turned out. The Hawaiians made no secret of their displeasure at seeing the ships back. They continued pilfering from the ships' store of fascinating objects. Cook was thoroughly familiar with the fact that Polynesians did not share English respect for private property, but he was short-tempered this time. Someone stole the *Resolution*'s cutter, a small, one-masted boat. Cook decided to take the Hawaiian king hostage until the cutter was returned. This tactic had usually worked to secure the return of stolen items, but it failed him now. When he went ashore on February 14, 1779, an angry crowd confronted him. It soon turned into a frenzied mob that moved forward and killed Cook and four marines in the ensuing scuffle. Attesting to the deep ambivalence Cook inspired, the Hawaiians then gave him the last rites of a revered chief.[23]

Journal entries from the crew of the *Resolution* and Hawaiian oral traditions provide the evidence for successive reconstructions of the nasty melee. Much turns on just exactly who the men and women of Hawaii thought Cook was. His comings and goings coincided with those of the ritual peregrinations of their fertility god, Lono, who was celebrated as long as the constellation Pleiades appeared in the sky—that is, from October to February. It is possible that when Cook returned to repair his ship, the period of Lono-veneration had been succeeded by that of a more malevolent being. What is indisputable is that the original veneration of Cook turned quickly into fury.[24] Back home, elegies, plays, paintings, and a French ballet competed to capture the high drama of this hero felled by raging savages. Cook's death almost overshadowed the accomplishments of his life.

In the preceding decade Cook had explored the unknown waters of the Pacific from Antarctica to the Arctic Circle. Now the whole habitable part of the globe had been probed. The search for the northwest passage had been laid to rest, at least until ice-cutters and global warming in the twentieth century made passage through it feasible for the first time. Cook had circumnavigated and mapped hundreds of islands and determined the location and dimensions of Australia after centuries of speculation. Coinciding with the ascendency of the British Empire, his feats announced the power of a fully modern nation ready to bend science and technology to its will.

In a dramatic coda to Cook's three voyages, in 1788 the British Admiralty sent William Bligh, now a commander, to Tahiti to test Banks's hypothesis that tropical plants could be grown anywhere in the same latitude. There was a utilitarian purpose to proving Banks right. Bligh was to carry to the Caribbean saplings of the amazing breadfruit that enabled Tahitians to subsist without work or worry. Bougainville's journal had first familiarized Europeans with this tree that provided ample nourishment eight months of the year. It was described as a "well tasted, wholesome, and nutritious" fruit, "the size of a boy's head," with no "part unprofitable except the rind which is thin."[25]

The botanist John Ellis called breadfruit a plant "bestowed by providence without the common tax of annual labor." "In the article of food," Banks remarked in his journal, "these happy people may almost be said to be exempt from the curse of our forefathers; scarcely can it be said that they earn their bread with the sweat of their brow when their cheifest sustenance Breadfruit is

procured with no more trouble than of climbing a tree and pull-
ing it down." The fact that a tree could produce nutrition over
eight months of the year undercut the biblical connection of labor
with sin and survival. Lucky Tahitians had no need to toil cultivating grains, pulses, or beans.[26]

Now the British government wanted Bligh to transplant Tahitian breadfruit to its West Indian sugar plantations where slaves
could only work and worry. Banks's gardeners outfitted Bligh's
ship, the *Bounty*, for transporting hundreds of saplings. Carrying seeds and plants across oceans was not easy. Hulls had to be
lined with copper to protect the plants from wood parasites, and
storms could easily damage such delicate cargo. In preparation for
the expedition, Bligh assembled six dozen shirts, fourteen dozen
looking glasses, eighty pounds of white, blue, and red glass beads,
twenty-two dozen stained-glass earring drops, 234 adzes, and
other ironware tools to pay for the trip's expenses.[27] But the *Bounty*
never got that far, for its crew mutinied against Bligh when it set
off for the Caribbean once the saplings were old enough to travel.

Contrary to literary tradition, Bligh was not a particularly
harsh commander. He shared Cook's concern for the health of his
sailors, even bringing along a fiddler to encourage the men's dancing for exercise on the deck. During the five months that the 1,015
seeded plants were growing, Bligh could have sailed the South
Pacific to map yet more islands, but instead he let the *Bounty* stay
in Tahiti. As a succession of Hollywood movies has graphically
depicted, that proved to be a colossal error of judgment. Most of
the crew got drunk every day, acquired tattoos, and took Tahitian
women as spouses. Ordered to leave, they obeyed, but the seduc-
tive power of their tropical home away from home soon manifested
itself. Bligh had "failed to anticipate," one contemporary noted,
"how his company would react to the severity and austerity of life
at sea after five dissolute hedonistic months at Tahiti." Within
four weeks of their departure, half of the forty-three-man crew
took command of the ship, forcing Bligh with eighteen loyal sea-

men into a launch. Bligh successfully steered the boat 3,600 miles to make landfall on the island of Timor.[28] It was a spectacular display of navigating skill. The mutineers had less luck—or skill.

In the ensuing court-martial, obligatory in the case of the loss of a ship, the admiralty exonerated Bligh of all charges and awarded him a prize for his navigational feat. They also let him finish his mission with new breadfruit saplings. In 1793 Bligh delivered the botanical treasures to Jamaica, where breadfruit became a popular food. Less fortunate, sixteen of the mutineers still in Tahiti were rounded up and sent home in chains. Eight other seamen followed Fletcher Christian, the mutiny's leader, to remote Pitcairn Island, where they torched the *Bounty* after stripping it of its wine and metals.[29] Despite their dependence on the hunting prowess of the six native men they had brought along, the mutineers abused them viciously and quarreled over the twelve women. The Tahitian men successfully plotted their revenge against the English, all but one of whom met a violent end within five years. When the British Navy finally reached Pitcairn Island, they found the one surviving seaman and twenty-four Anglo-Polynesian children. The mutineers had lived long enough to sire a nucleus of children who formed an enduring community.

The mutiny on the *Bounty*, like Cook's death, has inspired a succession of novels, plays, songs, and movies. It sped up a change of perspective on the Pacific. Over the course of fifteen British naval expeditions in the last third of the eighteenth century, the reputation of the Pacific changed from being the site of an Eden-like idyll to a place of violence and horror. In 1772 Maoris slaughtered and ate a group of French explorers who had been fishing off Tasmania, which seemed an isolated incident until Cook's murder. Increasingly the "noble savage" appeared less noble. Hearing how the Maoris had butchered the French, Rousseau asked with incredulity, "Is it possible that the good Children of Nature can really be so wicked?"[30]

At the end of the century, Great Britain was looking for someplace to transport its felons now that it had lost its continental American colonies. (They had jails for minor crimes and those awaiting trial, but prisons didn't become part of the criminal system until the nineteenth century.) Banks, whose love of Australia and botany never flagged, became a strong advocate of a settlement there. With his usual verve, he drew up a list of the plants that should be taken along: salad greens, beans, wheat, barley, oats, basil, fennel, marjoram, thyme, chives, mint, parsley, raspberry, gooseberry, strawberry, grape, orange, lemon, lime, apple, peach, nectarine, apricot, plum, and cherry. In 1788, the British founded a penal colony at Sydney.

Even poets became enamored of the scientific approach to nature. Oliver Goldsmith, known for his pastoral poem "The Deserted Village," wrote his eight-volume *History of the Earth and Animated Nature* from a naturalist's perspective. Biology now joined ships and weaponry to underpin conquests. Britain established links among its holdings in India, Australia, the Caribbean, and Canada. The goal became a profitable and self-sufficient empire.[31] Science had served the state well as a handmaiden in acquiring key possessions in the Pacific. Breadfruit from Tahiti could now feed slaves working in the sugar fields of the Caribbean. Maybe one day explorations would yield a northwest passage. Already British navigators had made the globe seem less a matter of landmasses and more a great ocean highway opening up the world's bounty to Europeans.

❋ CHAPTER EIGHT ❋

HUMBOLDT AND DARWIN
IN THE NEW WORLD

*Alexander Humboldt, the Prussian geographer, naturalist, and explorer,
with his guides at the Orinoco in 1800, woodcut after a painting by Ferdinand Keller.*

Amateurs—in the sense of being self-funded and self-directed—continued to play a critical role in creating knowledge, but something more was required to satisfy experts. Outcomes of investigations had to be precise, accurate, and open to inspection by others. This commitment to precision owed much to the perfecting of measuring devices such as quadrants, sextants, barometers, chronometers, microscopes, and telescopes. When Jonathan Swift fitted his hero, Dr. Gulliver, for a suit with a quadrant and compass, he was paying wry tribute to the new emphasis on instruments to facilitate the trained observer.[1] Where following the bent of one's mind used to be a preoccupation of gentlemen with leisure time, the new demands for accuracy narrowed the field to trained and dedicated practitioners.

Alexander Humboldt and Charles Darwin represented the finest in the tradition of independent inquiry and carried studies of nature to new levels of sophistication.[2] Although wide-ranging in his talents, Humboldt honed his skills, making definitive studies of the magnetism and topography of the globe's landmasses and the tides and currents of its oceans, tying together what had previously been studied separately. Darwin, working a generation later, wrestled with the one big and troubling question toward which dozens of inquiries had pointed: what were the origins of the earth's diverse species of flora, fauna, and humans?

As young men, both Humboldt and Darwin were dissatisfied with the conventional selection of careers offered them. In the end financial independence enabled each to follow his heart's desire and cultivate what turned out to be unique gifts. They spent years of travel and study in South America—Humboldt from

1799 to 1804; Darwin from 1831 to 1836—but their stellar accomplishments came from a peculiar genius that they shared, one that springs from insatiable curiosity and is buttressed by formidable powers of concentration.

Expeditions of several years in the Caribbean and Pacific taxed the courage and endurance of the most hardy. Whether commercial or naval, ships were small and crowded. They could be dashed on rocks and visited by mortal diseases. It's hard to exhaust the gamut of disasters possible on a long voyage: devastating fires, dwindling rations, fierce storms, fatal epidemics, and treacherous shoals, not to mention running aground. Then, upon arrival, there were the daunting tropical river basins of South America, teeming with animal and insect life, subject to extremes of weather, and home to various indigenous people, often hostile. The satisfaction of advancing knowledge didn't come to the fainthearted.

Though there was a scientific method by 1800, one that stressed accuracy and transparency, there were no scientists, or at least no one designated as a scientist. The term had not yet taken hold. Darwin in mid-century refers throughout his *Origin of Species* to naturalists. There were experts, people who specialized in a particular field such as botany or geology. If they held a university post, they rarely taught what their findings uncovered. Such research would be extracurricular activities. Charles Lyell, the pioneering geologist, is exemplary. He studied classics at Oxford and then became a lawyer, making geological observations in his free time. Only when his eyesight began to fail at age thirty did he devote his time exclusively to geology. Very few universities offered a full curriculum in the life sciences; support came, when it did, from museums, associations, and private collectors

Like so many naturalists before him, Humboldt spent his childhood amassing flowers, butterflies, beetles, shells, and stones gleaned from walks in the countryside outside Berlin, where he grew up. Retrospectively it seems prescient that at age eight, when visiting a different part of Germany, Humboldt asked, "Why don't the same things grow everywhere?" (Darwin made the same observation when he was ten.) Humboldt revealed an important truth about himself when he responded querulously to those who censured him for caring about too many things: "But can you really forbid a man from harboring a desire to know and embrace everything that surrounds him?"[3]

An avid reader, Humboldt drank in the accounts of the circumnavigation of the globe, including those by Louis Antoine de Bougainville, who returned to Europe the year of Humboldt's birth in 1769, and James Cook. The most critical influence in his life came from George Forster, who from age fourteen to seventeen had accompanied his father on Cook's second voyage. The Forsters were the first Germans to participate in exploratory voyages of the Pacific. George had written the naturalists' account, A Voyage Round the World, which Humboldt devoured as a boy.[4] Forster, sixteen years older than Humboldt, became a fast friend when the two met while Humboldt was a student at Göttingen and Forster was working on a scientific and literary journal published at the university there.

Forster taught Humboldt the importance of travel to science. The two went rock hunting in Holland and Belgium and then on to London, where Forster introduced Humboldt to Joseph Banks and other veterans of the Cook voyages. Humboldt got to inspect Banks's fabulous collection of botanical specimens and everything else that intrigued him. Forster was an excellent mentor, for he naturally integrated geological, climatic, cultural, and political data to gain an understanding of how human beings operated in their environment. When they were together, Humboldt and Forster visited peasants' houses, politicians' offices, museums, libraries, cathedrals, prisons, and observatories, as well as open fields, mines, wastelands, forests, and river valleys, carefully observing and recording their findings.[5]

Humboldt's mother had career plans for her son. Hired by the Prussian Department of Mines, he revived the ancient gold mines of the Fichtel Mountains. Indefatigable, Humboldt pored over sixteenth-century manuscript records of gold finds and then plunged into the mines for days on end to discover the seams of gold where the records said they would be. He installed a school for miners and devised new safety equipment, all the while conducting hundreds of experiments. He wrote these up in tracts on the salts of the Rhine, the diseases of miners, animal electricity, and the geography of plants.[6]

Poised for a brilliant career in the Prussian civil service, Humboldt elected to follow a different path, one that had opened up when he traveled with Forster. When his mother's death left him a wealthy man, he decided to devote the remainder of his life to studying nature in all its exuberant manifestations. Nothing that the Prussian government offered him could dissuade him from setting off to see the world on his own.

As a handsome Prussian baron, with a fortune to spend on his intellectual quests, the twenty-eight-year-old polymath arrived in Paris in 1798 like a meteor out of the East. But events can frustrate even charming aristocrats. Humboldt's first plan was to accompany an eccentric English nobleman bent on inspecting ancient ruins along the Nile. Napoleon's Egyptian campaign of 1799 quashed those plans, but another attractive proposition quickly materialized. Humboldt's boyhood hero, Bougainville, now seventy, invited him to join a state-financed, five-year expedition around the world with stay-overs in South America, the South Pole, and West Africa. Unfortunately, when war broke out again, the French government decided that it couldn't afford to go ahead with the voyage.

The postponed French mission did leave Humboldt with a great gift in the friendship of Aimé Bonpland, a ship's physician and talented botanist with a similar love of travel. In these waning years of the French Revolution, politics mattered too. The two men shared the same liberal sentiments. Humboldt and Bonpland agreed to explore the world's bounty together. While they were awaiting another chance to connect with an expedition, they went to Toulon, where they saw in the harbor the very ship Bougainville had sailed to Tahiti. They wangled an invitation to go on board. There Humboldt sought out that voyage's naturalist, Philibert Commerson, who had the cross-dressing valet.[7]

Tired of waiting for a formal expedition, the two friends took charge of their own travel destinations and set off on foot to Madrid. Humboldt had begun studying the earth's magnetism. Nightly he set up his equipment for celestial observations. The small-town Spaniards whom they encountered feared that this stranger was worshipping the moon, or worse, and often interrupted his astronomical observations. Still Humboldt managed to measure altitudes along the way with his new barometer. By the time he reached Madrid, he had determined that a hitherto unknown plateau extended across the interior of Spain.

Humboldt was establishing himself as a distinguished geographer; he already was a gifted linguist. When his connections arranged for him an interview with the king and queen of Spain, he aroused their cupidity with descriptions of the mineral riches he might discover in the New World. As the La Condamine expedition had already demonstrated, it was hard to travel in New Spain, but Humboldt managed to extract two passports from the monarchs to their Latin American empire.[8]

What seems surprising is that after many official Spanish explorations and the French Academy's geodesic expedition of the 1730s, little was still known about the interior of the South American continent. Setting sail from La Coruña, Spain, in the summer of 1799, Humboldt and Bonpland would correct this situation. During the next five years they carried the latest scientific equipment through tropics, rain forests, river basins, and up and down Andean peaks, covering almost 10,000 miles. Their brief stop in the Canary Islands on the way there had already disclosed Humboldt's penchant for seeing nature in its manifold relationships from the "heart soothing song" of the capirote to the "affinities" that might unite volcanoes like the one at Tenerife with Vesuvius and those in the Cordilleras of Peru and Mexico.[9]

Humboldt started using his instruments during the voyage. The faulty knowledge and ignorance about the New World exhilarated him. When the maps of the ship's pilot disagreed with one another, he used his instruments to clarify their position. Humboldt made his outbound voyage a feast of data collecting. Being well-equipped to examine almost anything that fell under his inquisitive gaze, he regularly measured the temperature of the seawater and the intensity of the magnetic force. He took frequent astronomical readings and used his cyanometer, an instrument designed to measure blueness, to observe the color of seawater. He also hauled in exotic fish with a net that he had hung from the boat's stern.[10]

The two explorers had actually planned to go to Cuba first, but a typhoid outbreak on board forced their ship to land in Cumaná, the site of Las Casas's failed settlement on the coast of Venezuela. One of the advantages of being fascinated by almost all natural phenomena is that changed plans could be taken in stride. Accidents and impulses would guide Humboldt and Bonpland for the rest of their five-year visit. Their focus remained the northwest corner of the South American continent.

While the ship was lying outside Cumaná, Humboldt, who spoke fluent Spanish, got the lay of the land from two Guaiquerie natives who canoed out to the ship to guide it into port. Columbus had called the area a second Garden of Eden. Humboldt was hardly less enthusiastic. He described a landscape filled with coconut palms, mimosa trees, whitish hills strewn with cactuses, and a bay filled with pelicans, egrets, and flamingoes, all under "a dazzling light." "The splendor of the day," he continued, "the vivid colouring of the vegetable world, the forms of the plants, the varied plumage of the birds, everything announced the grand aspect of nature in the equinoctial regions."[11]

Humboldt and Bonpland had not been in Venezuela long before they found an excuse to explore the interior. La Condomine, who had led the French Academy's expedition to measure the earth's latitude at the equator, reported being told of a natural canal that linked the Orinoco River with the Rio Negro, a part of the Amazon River system. Ever since La Condomine's return to Europe in 1744, European geographers had disputed its existence because there was no other such canal in the world. With his usual determination, Humboldt set off up the Orinoco to settle the issue. His group traveled for six months through tropics alive and noisy with biting insects, chattering monkeys, and birdsong. They spent more than two of those months in a dugout canoe packed to the gunwales with their servants, cooking utensils, and scientific equipment. Constant rain and tricky passages through cataracts made the trip miserable, but it proved worth it when they found the unique Casiquiare canal.[12]

Humboldt's curiosity extended from the laws governing tides and weather patterns to new botanical species and the manners, language, and history of the indigenous people he met in isolated areas of Venezuela, Columbia, and Ecuador. Being interested in so many things meant that he traveled heavy. Wherever he went, he carried sextants, a dipping needle, an instrument for calibrating magnetic variations, a thermometer, and one of Horace Bénedict Saussure's hygrometers, which used a human hair to measure humidity. While Humboldt let servants look after all this equipment, he never let his barometer out of his sight, and always carried a small sextant in his pocket.

Smarter than most of the people he met during his life, Humboldt was vain. He was also zealous to have his discoveries celebrated. When he visited Cuba after his excursion to find the Casiquiare canal, he was able to report that the location of Havana on contemporary maps was incorrect, a finding that earned him and Bonpland dinner with a grateful Spanish admiral. He loved solving mysteries like the existence of the Casiquiare canal, but even more, proving people wrong. Humboldt's passion for accuracy induced arrogance about his own work over the work of his predecessors. Given the steady improvement of instruments, it was probably unwarranted. La Condamine, long dead, earned a stern reproof. Appreciative of La Condamine's trailblazing through the Orinoco River basin, he still blamed him for circulating maps that were more mythical than cartographic.[13]

It's hard to get a grasp on Humboldt's prodigious curiosity. To detail the studies he made risks making him appear scattered when in fact he concentrated on grand physical laws and revealing relations among the parts of nature that had been subjected to scientific inquiry. All the phenomena that came before his eyes fit in as bits of evidence for larger pictures that suggested nature's harmony. But this broad swath of investigations should not suggest superficiality, for Humboldt was a very rigorous empiricist who insisted on careful notes and repetitive experiments. He could be called the first ecologist, for he strove to unite different inquiries as he tested their veracity. As he said upon leaving Europe, he wanted to "recognize the general connections that link organic beings." And he did, incessantly, by asking such questions as why the biting insects in Venezuelan rain forests differed from those along the coast of Colombia. Humboldt's celebration of union in diversity seemed particularly apt to Americans.[14]

Even while detailing his travels through the tropics with their lush vegetation, Humboldt stressed that he concentrated on the "intensity of magnetism or the strength of the earth's charge in different zones and elevations, as measured by the oscillations of a magnetized needle; hourly changes of the magnetic meridian; general meteorological phenomena; annual, monthly, and hourly mean decrease in temperature in the upper layers of the atmosphere . . . the regular ebb and flow of the aerial oceans, which in the tropics suffers no disturbance from weather changes. . . ." All this attention led to the discovery that the earth's magnetic intensity declines the farther one gets from the poles. He created labels and neologisms to pinpoint his findings, coming up with wonderful terms like "Jurassic" for the age of reptiles. But Humboldt's words do little justice to the imaginative way he responded to the sensual experience of taking in the pulsating life around him, even becoming overwhelmed by it.[15] He was a romantic figure in the era of romanticism.

Curious people kept running into each other; perhaps it would be more accurate to say that they sought each other out. Humboldt and Bonpland had hoped to join a French Arctic Circle voyage after their visit to Cuba. When they couldn't join that expedition, they decided to go to Bogotá to compare botanical notes with José Celestino Mutis, probably the world's finest botanist of his time. Mutis was the physician who had accompanied the viceroy of New Granada to his post and stayed forty years to complete a royal study of the flora of tropical, plateau, and mountainous areas. If those with similar concerns couldn't meet, their writings communicated for them. Humboldt encountered a man in the interior of Venezuela who had successfully crafted an electrical apparatus by reading Benjamin Franklin's *Autobiography*. Print knit the network of those with a scientific bent.

Mutis, thrilled at the prospect of a visit from Humboldt and Bonpland, dispatched a cavalcade of city fathers to greet the travelers. He put at their disposal the archbishop's London-made carriage. Behind the official welcoming party, a motley crowd of boys formed an informal escort stretching a quarter of a mile along the plain leading to Bogotá.[16] Mutis encouraged his guests to make use of the twenty thousand specimens in his botanical collection, which was second only to the one that Joseph Banks left to the Kew Royal Botanical Gardens. During their outings around Bogotá, Humboldt added a study of volcanoes to his fossil collecting, specimen gathering, astronomical sightings, and mineral searching. All of these activities, of course, contributed to his legacy in oceanography, meteorology, geology, and geography, not to mention the rich resource of his conjectures about the relations among all these phenomena.

Departing from Bogotá, Humboldt and Bonpland made their way to Quito, an arduous trip of four months that took them through striking contrasts of weather and terrain: the snow-covered Quindio Pass in the cordilleras with twenty-foot-deep ravines covered in vegetation so dense it blocked the sunlight, all topped off by the frozen mountain plateau just east of Quito. Both explorers lived their principles, refusing to be carried in sedan chairs. On the way they became experienced mountaineers. La Condamine had climbed part way up Mount Chimborazo, which rises over 20,000 feet above sea level. This provoked Humboldt and Bonpland to go even farther up the highest peak then known to Europeans. In Humboldt's account of the ascent, they followed a trail five to six inches in width winding upward between ice-encrusted snow on one side and shifting rocks on the other. At 15,000 feet, Bonpland found a butterfly. They were relieved of the company of flies at 16,600 feet. The climbers continued even while bleeding from their gums and lips, feeling nauseated, and suffering from vertigo, while their eyebrows and beards were encrusted with crystals of ice. (Twenty years later Simón Bolívar would claim to have ascended even higher, a feat laden with South American hubris.)[17]

Humboldt made sure that people back home knew of his successes. When word reached Paris that he had set a record by climbing almost 20,000 feet up Chimborazo, his reputation soared. This intrepidity earned him a singular popularity, even though a party of four had made this historic Andean climb. Some thirty years later, when botanist Joseph Dalton Hooker ascended one of the Himalayas, an aging Humboldt lamented, "all my life I have imagined that of all mortals I was the one who had risen highest in the world." He neglected the fact that Bonpland and two others had accompanied him.[18]

The Andean volcanoes rising straight up from the equator offered a man of Humboldt's inquisitive intellect a wonderful opportunity to observe the impact of altitude upon climate and vegetation. During his hikes he carefully noted the plants, insects, and animals appropriate to each height. When he returned home and studied his findings, he laid out six zones of altitude, each with its own ecology. His yearning for accurate but embracing laws of nature led Humboldt to these comparisons. Coining the phrase "isothermal lines," he used equal mean temperatures to establish different climatic areas, better known as temperate, tropical, and boreal vegetative zones. He had grasped that the vertical geography rising from desert floors to mountain peaks was akin to the horizontal geography across continents and oceans, an insight that enabled him to see weather as the great integrative force of the globe.[19]

There was yet one more accomplishment in South America to round out this extraordinary trip. Sailing north from Lima, Humboldt kept up the incessant readings of instruments that enabled him to mark the parameters of the current that flows northwestward from the southern tip of Chile to northern Peru. Although it was well known to the fishermen along the coast, Humboldt was the first to measure this large marine ecosystem, which supports the world's largest fishery. He calculated that surface currents carried away cold water, permitting nutrient-rich, warmer, upwelling water to come to the surface. Humboldt found that in some places the warm current extended 600 miles into the ocean. If climbing Chimborazo gave Humboldt popular acclaim in Europe, this study gave him a kind of Latin American immortality when the phenomenon was named the Humboldt Current.

Humboldt found more than the dimensions of the current off Peru. He was the first European to give serious attention to the Chincha Islands, where cormorants and pelicans had defecated for over a millennium. Digging into the 100-foot-deep piles of guano, Andean Indians had replenished their depleted soils since time immemorial. Smelling the islands a quarter of a mile out to sea, the ever-curious Humboldt carefully inspected this natural treasure. He carved out several samples to be sent to two French chemists. The stench repelled shippers for another generation, when intensive farming and a surer grasp of guano's fertilizing potential sent enterprising traders to the dozens of guano islands that lay 500 miles off the coast of Peru. The ensuing exploitation of Africans and Chinese in one of the world's worst slave regimes would have appalled Humboldt.[20]

Humboldt spent the last year of his expedition in Mexico, which he visited with the express intention of writing a study that embodied the geography, economy, and politics of an entire country, in this case the most advanced entity in the Spanish Empire. George Forster's influence is evident in the statistics extracted from offices and libraries. Humboldt drew on his excursions climbing the country's volcanoes and correcting the maps of the west coast. His earlier expertise gained as a Prussian inspector of mines came into play when he examined the great silver deposits at Taxco and Guanajuato. A severe critic of the Spanish colonial administration, Humboldt considered the Mexicans capable of self-government.[21]

Although Humboldt was eager to get back to Paris with his treasured material, he was not content to leave the New World without visiting the new nation to the north, which held out so much promise for realizing his political ideals. He and Bonpland sailed to Philadelphia from Mexico in the spring of 1804. The

United States embodied all the political principles of these two liberal Europeans, except for its slavery, which they had had many opportunities to scorn on their travels.

In Philadelphia, Humboldt and Bonpland met the city's intellectual luminaries—Benjamin Rush, Benjamin Barton, Caspar Wistar, and Charles Willson Peale. Humboldt, as usual, dazzled them all. Leafing through a scientific journal at Peale's "omnium and gatherum" museum, he let out a whoop of joy. A news story in the journal reported that the manuscripts and specimens he had sent to his brother had arrived safely.[22] After accompanying Humboldt on a ride to Washington, Peale described the German baron as speaking continually in a rapid mixture of English, French, and Spanish, adding that he was "very communicative and possesses a surprising fund of knowledge in botany, mineralogy, astronomy, philosophy, and natural history." [23] The purpose of this final leg of their voyage was to meet the president, Thomas Jefferson, who had just dispatched Meriwether Lewis and William Clark with the thirty-one-member Corps of Discovery to explore the vast area he had just purchased from Napoleon.

Jefferson was eager to learn as much as possible about the terrain where the United States now abutted Mexico. Humboldt displayed his usual generosity, lending him maps and notes filled with statistics. Albert Gallatin, the Swiss-born secretary of the treasury, rendered another account of Humboldt's conversation. He described to his wife the rapid-fire speech of Humboldt but allowed that he had been "really delighted, and swallowed more information of various kinds in less than two hours than I had for the two years past in all I had read or heard."[24]

Humboldt and Bonpland embarked homeward from Philadelphia in June 1804 with forty cases of fossils, specimens, equipment, maps, and notebooks. Humboldt had spent a third of his fortune on the trip, but, as he exulted, his sphere of ideas had been "wonderfully enlarged." Presciently, he saw that the Europeans' wholesale destruction of the forests in the Western Hemisphere could actually affect climate. Very little in the natural world escaped Humboldt's discerning eye.

Three months later Humboldt opened his botanical collection of some sixty thousand specimens—many previously unknown— to the public in a display at the Jardin des Plantes. Humboldt himself was in great demand. Public attention turned him into a celebrity, a status he never lost in the remaining fifty-five years of his life. People treated him like a hero. He addressed groups large and small, from a packed audience at the Institut National to the famous salon of Mme de Staël and the country house at Arcueil where the French chemist comte Claude-Louis Berthollet entertained the intellectual elite of Paris. Only Napoleon, no doubt jealous of the fuss made over Humboldt, seemed impervious to Humboldt's intellect or charm. After inviting him to the Tuileries, where he was about to declare himself emperor, he condescendingly compared Humboldt's botanical travels to his wife's plant collecting.[25]

Humboldt had no need of the emperor's praise. He was well aware of himself as an innovator. Before settling down in Paris, Humboldt went to Rome to visit his brother, who was the Prussian ambassador to the Vatican. There he met a young Venezuelan named Bolívar, who sought his opinion about the readiness for self-government among the creoles in the Spanish Empire. Whether justified or not, Latin Americans give Humboldt credit for encouraging Bolívar, who led successful wars of liberation in Colombia, Venezuela, Ecuador, Peru, Panama, and Bolivia in the next decade. For Humboldt these were bittersweet victories, for Bolívar's forces often executed the liberal aristocrats who had entertained him and given him expectations for peaceful reform.

In the case of Mexico, Humboldt's influence on politics was more direct. His *Political Essay on the Kingdom of New Spain* provided concrete details about Mexico's natural resources accompanied by precise recommendations for how the economy might sustain an independent nation. For this accomplishment, the newly independent Mexican government bestowed honorary citizenship on both Humboldt and Bonpland.[26]

Returning to Paris after visiting his brother, Humboldt set to working up his material for publication. Here was a project! It involved directing the labors of cartographers refining his maps and astronomers reducing his astronomical observations. When Bonpland proved surprisingly unwilling or unable to prepare thousands of specimens for publication, Humboldt hired a botanist to complete the job. One by one, the thirty volumes of Humboldt's *Voyage de Humboldt et Bonpland* appeared between 1807 and 1834. They opened up for Europeans the fabulous tropics and mountain ranges of South America, long closed off from them by the secrecy of Spanish authorities that had become policy after the first generation of Oviedo, Cortés, Columbus, Pigafetta, and Las Casas. With its 1,425 maps and plates, Humboldt's volumes drained what was left of his fortune. The volumes were so costly that he couldn't afford a complete set for himself. Only Napoleon's *Description de l'Egypte* cost more.[27]

Even this great undertaking could not absorb all of Humboldt's abounding energy. While lecturing and socializing widely, Humboldt turned out several other studies, including *Aspects of Nature*, a lovely evocation of nature's beauty to be found in Latin American landscapes. This became his most popular work.[28] During his years in Paris he befriended many a young man embarking on a career in science, from Justus Liebig, the pioneer of organic chemistry, and the mathematical genius Carl Friedrich Gauss to Adolphe Quetelet, who applied the model of astronomy to social studies and created modern statistics. Louis Agassiz, the paleontologist, was another friend, as was the young geologist Charles Lyell, whom Humboldt helped date fossils.

Humboldt wrote between one and two thousand letters a year,

bringing together fellow investigators who would otherwise not know each other. He dispatched letters to people who might contribute funds to sustain his expensive publishing projects. In later years his annual correspondence rose to three thousand letters. The introduction of transatlantic steamers brought new clusters of visitors to his door. Humboldt gave a boost to international scientific collaboration, too, when he successfully pushed for a chain of geomagnetic observation stations across the globe. He organized the first international conference of scientists, which was held in Berlin.

With impressive determination Humboldt had galvanized his powers of observation and directed them toward the mother lode of knowledge to be gained through measurement. He greatly augmented the categorizations of the Linnaeans and drew future investigators to what one scholar memorably described as "nature in motion, powered by life forces many of which are invisible to the human eye; a nature that dwarfs humans, commands their being, arouses their passions, defies their power of perception."[29]

Humboldt had returned to a Europe torn to pieces by Napoleon, who had carried his reforming impulses along with his army as it marched across the continent. Napoleon put an end to the Holy Roman Empire, which had offered, since 800, a kind of institutional cohesion to Prussia, Austria, and some three hundred principalities and cities that composed the German-speaking world. The Napoleonic wars also delivered the coup de grâce to Humboldt's already diminished fortune, but he did acquire a patron. As an undergraduate at Göttingen, Humboldt had shared a room with Klemens von Metternich, who as Austrian foreign minister came to Paris in 1814 to exult over the defeat of Napoleon. Metternich then drafted Humboldt to escort Prussia's King Frederick William III around town.[30] Enchanted by the voluble Humboldt, the king gave him a stipend as royal chamberlain. After that, Frederick William allowed Humboldt to stay in his beloved Paris until 1827, when he finally summoned him back to Berlin to perform his duties at the Prussian court.

Coming back to Prussia exposed Humboldt to the reigning philosophy at Berlin University where scholars approached nature with intuition, awe, and wonder. To combat this indifference to empirical work, Humboldt began a course of lectures, starting at the end of 1827, on physical geography. He delivered them at a lecture hall in the Berlin University, which his brother had founded seventeen years earlier. Speaking extemporaneously, he enchanted audiences of students, professors, and thousands of the general public. He talked about volcanoes, sunspots, meteorites, optics, the motions of planets, and the distribution of plants and animals on the earth's surface, weaving all the information into a complex pattern. He gloried in popularizing science, but perhaps even more important, he knew he was introducing his audience to a freedom of inquiry about nature that was denied them politically in authoritarian Prussia. After repeat performances of his lectures, Humboldt devoted the next thirty years to publishing these lectures, the last installment going to the printers in the year of his death, 1859.

Entitled *Cosmos*, the five-volume work has been called the "last great work of the last great universal man." Humboldt's goal was to demonstrate the connectedness of the world, to embrace the missing links of knowledge rather than sacrifice them to an incomplete portrayal of how nature worked. It turned out to be amazingly popular, becoming the educated person's introduction to nature's force and beauty as well as the state of scientific knowledge in mid-century. In it, Humboldt continued his struggle to find unity and harmony in the multifarious manifestations of nature's power. To this end he included a long section on language in tribute to his brother's interest in philology, but also to express the hope that through communication "the barriers which prejudice and limited views of every kind have erected among men" might be erased in deference to treating "all mankind, without reference to religion, nation, or color, as one fraternity, one great community."[31]

Humboldt was thirty-five when he returned from the New World. Through the next half century, medals, honors, academic society memberships, and royal accolades rained down on him. Wherever he went there were hundreds ready to sit at his feet and listen—and he loved to talk about what he had done and what he knew. The Royal Society, bestowing on Humboldt its highest honor, noted the vastness of the facts that he had given to the world that "will be regarded with astonishment as the work of one man." What astounds today is how little Humboldt is remembered. Contemporaries said he was second in fame only to Napoleon, born the same year. A succession of scientists attested to his critical importance to their careers and their fields. "Every scientist is a descendent of Humboldt," one wrote, adding, "we are all his family."[32] Few great men have had their reputations fade so quickly, just leaving his name on a bay, peak, lake, current, sinkhole, penguin, lily, orchid, and oak whose namesake few remember.[33]

A latter-day Bacon, Humboldt helped drive the last vestiges of metaphysics from the study of nature. He shared Bacon's disdain for opinion, though for him untested theory became the target of his scorn. Both men had visions about how knowledge might create a better world, using the leverage of facts to push aside the superstition that ignorance spawned and the moribund institutions that dressed up ignorance as tradition. Humboldt, arriving on the scene two centuries after Bacon, nurtured more liberal, egalitarian hopes for his fellow human beings as he strove to save the beauty of nature without sacrificing facts to sentimentality. Like Bacon he wanted to infuse those about him with the intellectual excitement that had fueled his life's work.

The adulation of Humboldt marks a particular moment in

the West's reception of science. To compare his writings with the entries in Diderot and Jean Le Rond d'Alembert's *Encyclopédie* is to traverse the ground between interesting accounts of experiments and powerful explanations of natural forces. The mastery that Newton's work had conveyed for the solar system, Humboldt's work crystallized for plant geology and geophysics. Working on ecology before anyone coined the term, Humboldt supplied the methods for addressing the tantalizing relations of organisms with their physical environment.[34]

While contemporaries marveled at all the new facts that Humboldt had gathered, those who followed in his tracks admired more the light that his imagination cast on the path ahead. By the time of his death in 1859, the hopes of the Enlightenment had turned into the confidence in progress that marked the nineteenth century. Yet science was still a largely random pursuit of truths dependent upon the curiosity of pioneer practitioners. It was orderly, but not well organized. The life of the great unifier of biological studies, Charles Darwin, underlines this fact. One might wonder if Darwin would have become a naturalist at all without the marvelous books Humboldt wrote and he enthusiastically consumed.

When Darwin turned sixteen, science was taught in many universities, German and Scottish ones in particular, but it hardly constituted a career choice. Indeed, the notion of a career was novel. The word acquired a new connotation in Darwin's lifetime, when replicating the path of one's parents no longer described what many men—and even some women—were actually doing. Where formerly career referred to "the ground on which a race is run," in the first decade of the nineteenth century people started using it to talk about "a person's . . . progress through life."[35]

Darwin's father, a physician, hoped that Charles would become a doctor. He sent him to Edinburgh where his older son was already studying medicine. In retrospect, Darwin found medicine unbearably dull. He considered the years at university in Scotland, from 1825 to 1827, a waste. He recalled a single intellectual gain. While out walking with a fellow student, he learned about Jean-Baptiste Lamarck's views on evolution. His letters reveal other more relevant experiences than his grim, retrospective judgment suggests. He had gone on geological excursions, studied taxidermy, and collected seashells on the shores of the Firth of Forth.

Frustrated, but not deflected, Darwin's father next dispatched Charles to Cambridge to become a clergyman in the Church of England. There, Darwin reflected, "my time was wasted, as far as the academical studies were concerned, as completely as at Edinburgh." He did make a thorough acquaintance with William Paley's *Evidences of Christianity*, a very popular work building on John Ray's demonstration of the existence of God from the design in nature, especially the perfection of the human body.[36]

In these days Darwin was something of a sport, hanging out with the young blades of Cambridge. "I should have thought myself mad to give up the first days of partridge-shooting for geology or any other science," he wrote in an autobiographical sketch. But he did make friends with two eminent naturalists. John Henslow, the Regius Professor of Botany, held weekly open houses for undergraduates interested in scientific topics. Darwin became so attentive to him that he became known as "the man who walks with Henslow." Another don, Adam Sedgwick, took Darwin along for a geological research trip to North Wales. Astonished at finding a tropical shell there, Darwin mused that he had never before realized that "science consists in grouping facts so that general laws or conclusions may be drawn from them."[37]

Henslow and Sedgwick were to remain important figures in Darwin's life. Henslow discerned Darwin's striking gifts as an investigator of nature. He opened the door to a different calling for him when he recommended him for the position of naturalist on a survey voyage of H.M.S. *Beagle*.[38] Two of Darwin's future allies, the comparative anatomist Thomas Huxley and the botanist Joseph Hooker, had launched their own careers on similar naval expeditions.

In his autobiography Darwin commented that work in mathematics had been repugnant to him.[39] That would prove to be a decisive aversion. Mathematics since Galileo and Newton had given the physical sciences the kind of proof that rendered them superior to the life sciences. Darwin was to discover what Buffon had stressed a century earlier: botany, zoology, ornithology, and geology required getting outdoors and closely examining natural phenomena. These observational skills were as appropriate for the study of living organisms as mathematical calculations were to astronomers and physicists. With these talents Darwin was able to escape his father's career planning. He sailed from England on the *Beagle* in December 1831, not to return for almost five years. Aside from some geology and a stint of beetle collecting, Darwin had little preparation for the job ahead.

The reluctance of Darwin's father to support his son's trip on the *Beagle* is a bit puzzling because his own father, Erasmus Darwin, had achieved fame as a natural philosopher, even hinting at the theory of evolution in his long poems about flora. His poem *The Love of the Plants* popularized Linnaeus's system of classification in England and his personification of the plants in the Linnaean taxonomy in amusing sketches made quite a hit as well as a bit of notoriety for him.[40] But he had practiced medicine and written his books as a sedentary speculator. His grandson, like Humboldt, eschewed this course of action, choosing instead to become the intrepid investigator, traveling great distances to glimpse the scope of nature's proliferation. These trips with their invitation to general speculations also distinguished Darwin from the experimental zoologists of their day who studied the functioning of animals in laboratory experiments.

Darwin carried Humboldt's *Personal Narrative of Travels to the Equinoctial Regions* aboard the *Beagle*. It detailed the people, plants, animals, climates, and topographical features encountered in the New World. A somewhat disorganized affair, the *Personal Narrative* managed to convey both Humboldt's passion for nature and his obsession with accuracy. The five-volume English translation appeared in 1825, when Darwin was sixteen. "My whole course of life," he subsequently wrote of Humboldt, "is due to having read and re-read his Personal Narrative as a youth." On the *Beagle* he admitted to his diary "I formerly admired Humboldt, now I almost adore him."[41]

Humboldt had managed to convey the exhilaration of watching closely as bugs, flowers, birds, and animals pursued their survival strategies with such writing as this: "What a fabulous and extravagant country we're in . . . fantastic plants, electric eels, armadillos, monkeys, parrots: and many, many, real, half-savage Indians. What trees! Coconut palms, 50 to 60 feet high; Poinciana pulcherrima with a big bouquet of wonderful crimson flower, pisang and a whole host of trees with enormous leaves and sweet smelling flowers as big as your hand, all utterly new to us. As for the colouring of the birds and fishes—even the crabs are sky-blue and yellow."[42]

Darwin called Humboldt the parent of "a grand progeny of scientific travellers" and would mention him four hundred times in his writing. It would be hard to exaggerate the impact of America's tropical splendor upon Europeans with an inquisitive bent. When Darwin arrived in South America, he announced that his mind was "a chaos of delight. . . . I am at present fit only to read Humboldt; he like another sun illumines everything I behold." Looking back more than forty years later, Darwin had succumbed to the New World as much as Humboldt had. He invoked "the glories of the vegetation of the Tropics" that rose before his mind, adding that "the great deserts of Patagonia and the forest-clad mountains of Tierra del Fuego . . . left an indelible impression on my mind."[43] Darwin, like Oviedo, Mutis, and Humboldt, became utterly entranced by the sights and smells of the New World.

Eighteen of the fifty-seven months of the *Beagle*'s voyage were spent at sea and thirty-nine on land. The ship had been outfitted for a hydrographic survey to study the shorelines, tides, and currents, as well as submerged objects that could affect maritime navigation. Its commander, Robert Fitzroy, who was four years older than the twenty-two-year-old Darwin, shared a cabin with Darwin and directed him to collect natural specimens. Darwin proved good at this. He found in Patagonia's cliffs fossils from huge mammals cheek by jowl with seashells. Although the ship visited Brazil, the Falkland Islands, Argentina, Chile, Peru, Tahiti, New Zealand, Australia, and Mauritius, the five weeks spent in the Galápagos archipelago proved the most fruitful for Darwin's research because their closeness to, and separation from, South America, had created a unique environment. The mix of similarity and difference was the evidence that pushed Darwin's toward a theory of gradual speciation.

Six hundred miles west of Ecuador, the eighteen volcanic islands in the Galápagos archipelago straddle the equator. The *Beagle* arrived among them in September 1835. Darwin went from one dazzling observation to the next. He carefully detailed the burrowing home building of lizards along with the characteristics of the Compositae family of daisies and asters that tied the Galapagos Islands to the South American continent rather than to the Pacific. What struck him with wonder was that the various islands, "most of them in sight of each other," were "differently tenanted." As he explained, "several of the islands possess their own species of the tortoise, mocking-thrush, finches, and numerous plants . . . occupying analogous situations, and obviously filling the same place in the natural economy of this archipelago. . . ." Darwin's observations through the many stops made during the

Beagle's circumnavigation of the globe strengthened his suspicions that species might not be stable.[44]

Among those working on scientific topics, there was already a desire to find a natural explanation for the origin of new species. Humboldt's whole career suggested that quantification applied to nature through measurements could lead to breakthroughs when it was leavened by an investigative spirit. An aesthetic appreciation played a big part in Humboldt's approach to nature, even though his legacy included the conviction that with reason the core of reality could be grasped. John Herschel, a noted mathematician and astronomer of Humboldt's generation, urged his fellow naturalists to search for universal laws when studying natural phenomena. Darwin met Herschel when the *Beagle* stopped at Cape Town on its way back to England. He had taken in Herschel's point; he wrote in his journal, "if data is to be useful it must be gathered for or against any idea." He now had his idea.[45]

It's hard not to perceive Darwin's time on the Galapagos with the illumination of hindsight. Yet Darwin himself called his visit the most critical experience of his life. "Reviewing the facts here given, one is astonished at the amount of creative force," he commented, "if such an expression may be used, displayed on these small, barren, and rocky islands." There were other more immediate and personal consequences of his voyage on the Beagle. Toward the end of the trip, Darwin learned that his old professor, Sedgwick, after reading some of the letters he sent home, told his father that Charles would "take a place among the leading scientific men." This validation stirred a fierce ambition in Darwin. It also calmed his father's anxiety about his future, always a relief to a dutiful son.[46]

Drawing on his letters, notes, and journal entries, Darwin composed the naturalist's record for the second volume of Captain Fitzroy's three-volume report. A sharp-eyed publisher saw the potential in Darwin's piece, plucked it from the official document, and published a pirated edition in 1845. This brought Darwin fame and the attention of England's most prominent men of science, including Charles Lyell and Thomas Huxley, with whom he corresponded for the rest of his life.[47]

Since Darwin's work, the word "evolution" brings to mind inherited variations, the struggle for existence, and nature's selection of the most fit, but the idea of evolution had a substantial history before Darwin took up the subject. His grandfather had hinted at it in his *Zoonomia*. Denis Diderot, not an expert, but a remarkably curious man, had raised the idea of possible changes in species. Lamarck, Buffon's most gifted student, hypothesized that living organisms changed over time by inheriting acquired and beneficial characteristics in his *Hydrogeology* of 1802, the first systematic account of the history of natural phenomena. At the time, Lamarck's insistence on the continuity between humans and animals was an even more important insight for biologists.

Here Lamarck followed Linnaeus, who had told the Lutheran archbishop of Uppsala that he knew no way to disguise the similarity of humans to simians. Given the centrality of the biblical story of how God created the earth and its flora and fauna for human beings, taking this essential first step marked a prime intellectual advance. Linnaeus had been uncertain that species changed over time because he had difficulty explaining why the Almighty would have created such a proliferation of species. Far more aggressively, Lamarck refused to place any investigation off-limits on religious grounds. On a different front, Lamarck's striking illumination of the hidden processes of change in nature helped drive out the Cartesian view that organisms were merely complex machines.[48] While daring in its implications, Lamarck's central idea of improvement sat well with the nineteenth-century infatuation with the progress of civilization.

Humboldt too embraced evolution. He couldn't explain the origin of species, even though he demonstrated that human life is embedded in a specific climate and affected by particular vegetation, the whole varying with different altitudes and soil conditions. He attributed the diversity among races to processes of adapting to different environments, and he relied on fossils to date geological strata. Yet the cause of evolution—the mechanism for change—remained a mystery to him; he simply took it for granted. Humboldt's ruminations on the origins of mountain ranges and the relation of earthquakes to the earth's crust fed many of Darwin's early speculations about evolution. Even more importantly, the field of plant biography, which Humboldt had really invented, helped Darwin understand the distribution of species.[49] Yet the study of nature that led Humboldt to unity and harmony took a darker turn in Darwin's depiction of the inexorable struggle for existence.

How acquired characteristics might be transmitted to the next generation represented the biggest hurdle Lamarckians had to clear. Yet the idea possessed a plausibility that helps explain why Darwin continued to consider this possibility. He never rejected Lamarck completely, even though his theory moved in a different direction. Darwin said that change came gradually, as inheritable variations of specific traits enabled their carriers to adapt better to their environment, not, as Lamarck thought, that qualities contributing to survival were acquired and passed on.

The possibility of random sources of variation did not enter the store of biological knowledge until the early twentieth century, when the work of the Austrian Augustinian friar Gregor Mendel became widely known. Mendel experimented with twenty-nine thousand pea plants for a decade before publicizing his findings about inheritance patterns. A classic example of finding the wrong audience, Mendel's 1865 paper reached those concerned with hybridization, not inheritance. It fell to Thomas Hunt Morgan to integrate Mendel's research into the chromosome theory of inheritance to form the core of modern genetics.[50]

To take up the trail of influences playing upon Darwin's imagination is to revisit the enduring fascination with fossils, those awkward reminders that some of the earth's original species had disappeared. A trio of seventeenth-century naturalists converged on the idea that the solid bodies within rocks were actually remains of once-living organisms. Studying the teeth of a recently killed shark prompted Nicolaus Steno in 1667 to conjecture that the embedded material of fossils was initially liquid. He concluded that the strata in rock formation had once been horizontal to the earth, making its current placement a marker of geological time.[51]

John Ray, whom we met as the clergyman and predecessor of Linnaeus in forming a taxonomy of living forms, also contributed to evolutionary thinking. He used his knowledge to promote the idea that the study of nature is an appropriate form of worship. Although Ray interpreted the diverse ways that plants adapted to different environments as evidence of divinity, his close observations of the relation of forms to functions in living things opened a new way of thinking.[52] In many ways he is the forerunner of today's believers in Intelligent Design.

Georges Cuvier, Lamarck's younger colleague at the Jardin du Roi Museum in Paris, had been inspired by Buffon's works, principally to challenge his many conclusions as well as Lamarck's endorsement of progress. Fascinated by paleontology, he dug up mammal fossils in the middle of Paris! (Fossils had turned up centuries earlier when the Romans were mining for gypsum.) Cuvier examined the neglected remains of bears, elephants, deer, and rhinoceroses. Honoré de Balzac wrote ecstatically of Cuvier, "he digs out fragments of gypsum, decries a footprint and cries out: 'Behold!' and suddenly marble turns into animals, dead things live anew, and lost worlds are unfolded before us!" Cuvier demonstrated that the animal world, existing before humans appeared, could be reconstructed.[53]

Cuvier compared fossil skeletons of Indian and African elephants with American mammoths and determined that mammoths were different and hence an extinct species. From that
conclusion he found other extinct species, a telling discovery in
the last decade of the eighteenth century when almost everyone
believed that God's perfection as a creator negated the possibility of an extinct species. Buffon had argued that God didn't
make rough drafts and that fossil remains indicated that species
had moved out of the areas where their skeletons were found.
Cuvier's work demonstrated the falseness of Buffon's explanation.
His comparisons demanded close comparison of living species
with extinct ones, which he accomplished with rare skill.

Cuvier noted that there had been more elephant fossils found
in Germany than in any other place. This was not, he explained
wryly, because there had been more elephants there, but rather
because there were more people throughout Germanic lands
capable of identifying them. Even the great poet Goethe took up
paleontology, remarking fifty years before Darwin on the similarity of the human jawbone to the ape's.[54] Cuvier was the first
to speculate that there had been an era when giant lizards had
been the dominant animals on earth. The theory of evolution was
acquiring weight. It was moving beyond its role as plug for the
hole of ignorance about fossils, species, and the age of the earth,
but it would take an adroit arrangement of evidence to counteract
its offensiveness to most people, naturalists included.

Passing well beyond Linnaeus's concern with naming and
classifying living things, Cuvier figured out how the parts of the
invertebrates he studied functioned. He detailed their interaction
in mollusks, crustaceans, insects, and worms in a way never done
before, giving his attention to the organs of nutrition, then circulation, and finally nervous systems. After these investigations
he was able to explain the correlation of parts. The evidence of
extinct species in itself didn't predispose him to the idea of evolu-

tion. Still his fine comparisons among species made it much easier for Darwin to come up with his explanation.[55]

Darwin spent five years in London after the return from his voyage in 1836. The first two years were crucial, for he studied geology under the tutelage of Charles Lyell. He had actually bought the first volume of Lyell's *Principles of Geology* in Montevideo while on the *Beagle*. His recent experiences had pushed him away from Cuvier's and Sedgwick's teachings that catastrophes such as episodic flooding had dominated geological processes and toward Lyell's view that change came from the aggregation of the effects of natural processes on flora and fauna operating over a long period of time. The notion of extended periods of time opened up the possibility of change so gradual that it could hardly be perceived yet still be effective in changing species.

Everything that Darwin read nurtured his imagination during this intellectually charged period. In Thomas Malthus's *Essay on the Principle of Population* he encountered the idea that living organisms reproduced exponentially while their food supply grew arithmetically, if at all. This thesis underpinned Malthus's rejection of the optimistic expectation of future progress. In such a situation a portion of human progeny must die through what he categorized as the positive checks of starvation, disease, and predation. Malthus provided a theory for the scramble to survive that Darwin had already observed in extensive observations of animal life. "Being well prepared to appreciate the struggle for existence which everywhere goes on," Darwin reflected later, "it at once struck me that under these circumstances favourable variation would tend to be preserved, and unfavourable ones destroyed."[56]

Malthus's analysis acted as a catalyst in Darwin's mix of ideas. Adam Smith's *Wealth of Nations* gave Darwin another push toward unlocking nature's secrets, especially Smith's conjecture that individual competition could lead to social betterment through the operation of unintended consequences. His notes from this period chronicle his thrashing about with fresh suggestions, abandoned hypotheses, and conceptual dead ends, but his concentration never flagged.

Darwin apparently never thought of creating a great private collection like Hans Sloan or Joseph Banks. Instead he sent his specimens to various British institutions. And he profited from this. When he donated his mammal and bird specimens to the Geological Society of London in 1837, the ornithologist John Gould reported back that twenty-five of Darwin's land birds were new and distinct species found nowhere else in the world. Yet, most important, they all had close affinities to finches and mockingbirds from the continent.

Island species geographically cut off from their continental origin differed little. The crisp lines of demarcation that essentialists insisted upon—each species endowed with an essence—didn't exist. Looking at the evidence, Darwin concluded that "species gradually become modified," adding rather ominously that "the subject has haunted me." Quickly he moved toward a theory that focused on the formation of new species through a steady accumulation of slight and propitious differences between parents and some of the offspring.[57] Here he was rejecting the idea of saltation, an abrupt evolutionary change from a sudden mutation. The operative term then had been transmutation, not evolution.

In early 1840, Darwin wed his cousin, Emma Wedgwood. Two years later, they left London to settle down to a life of solitude in his country home in Kent. Very happily married, he presided over a brood of children, seven of whom survived the fevers that killed two sons and a daughter. Darwin continued to study animal husbandry and experiment with the reproduction of plants. He wanted to learn if eggs and seed could safely move from one continent to another. He tested his theory about species on a large collection of pigeons and by conducting a voluminous correspondence. This was not idle letter writing but outreaches to civil servants, army officers, diplomats, fur trappers, horse breeders, hill farmers, society ladies, kennel hands, zookeepers, pigeon fanciers, gardeners, and asylum owners who could answer one of his many questions, posed as he was tying up the loose ends of his theory. For all his genius, he was never able to figure out the origins of flowers—"an abominable mystery," he said—but he strewed them across his estate.[58] He also suffered from ill-defined but frequent and painful ailments that only got worse with age.

Before leaving London for his country home Darwin had published *The Structure and Distribution of Coral Reefs*, in which he argued that coral reefs owed their existence to slow deposits of coral on volcanic cones.[59] These observations and reflections, published in 1842, had led him ineluctably to a materialist explanation of the origins of species. He had also acted as general editor for a series of publications under the title of *Zoology of the Voyage of H.M.S. Beagle*. His own *Voyage of the Beagle* had appeared in 1839. He didn't have Humboldt's passion for publishing; aside from a piece on barnacles, this was Darwin's output before his one big book.

It took Darwin until 1858 before he was ready to give the public his explanation of how species formed and reformed, even though he wrote the core of his argument in 1844. He hadn't frittered away his hours. He spent eight years studying barnacles while puzzling over his grand thesis. He knew from the scorn heaped on earlier writings about evolution just how fierce would be opposition to such a position. Deciding upon a course of action to avoid opprobrium became a grueling ordeal, intensified by moral and psychological pressures.

Fashioning a strategy for public presentation involved thinking through the positions of likely opponents. It took many years for Darwin to locate the error in the essentialist conviction to which Cuvier had remained true. The assumption that each life form had an essence didn't work, Darwin saw, because there was no one matrix from which all the models issued. Instead, natural phenomena had to be treated as a population of individual specimens. Living organisms had experienced billions of years of historic development that made them complex in a way unknown to inanimate objects. Even more significantly, each living thing was an individual. Whether a woman, a dog, a horse, or an amoeba, each member of a species had qualities that distinguished it from others (think of members of a family or a litter of cats). With this realization came the awareness of the enormous diversity in the natural world and the possibility that this diversity might divulge clues about how new species had emerged through heritable variations. "Nature" would favor those progeny with a variation that enhanced the capacity to survive in Malthus's world of food shortages.

To persuade readers of the evolution of species through the radical hypothesis called "natural selection" was going to take powerful analyses strengthened by irrefutable evidence. At least Darwin thought so, and he spent the years between 1844 and 1858 clarifying concepts and working out the intellectual conundrums that remained. Only rigor would save him from the ridicule his scientific colleagues heaped on the tentative propositions about evolution in Robert Chambers's *Vestiges of the Natural History of Creation*. Published in 1844, when Darwin was creating a draft of his new theory, *Vestiges* found admirers, running through eleven editions in the next twenty years. With a devout wife, Darwin was acutely aware of how his ideas would be an affront to many religious people. His notebooks allude to fear of attacks, instancing the "persecution of early Astronomers." As the father of a large family, he moved cautiously, hoping to minimize any alienation from the public.[60]

Darwin grasped the challenge to certain entrenched ideas. As he astutely put it, "anyone whose disposition leads him to attach more weight to unexplained difficulties than to the explanation of a certain number of facts will certainly reject my theory." His audience, he was saying, would be divided between those who saw the value of a sound but imperfect theory and those who would resist until every one of their doubts had been resolved. He counted on "young and rising naturalists," and they didn't let him down, although there was enough ambiguity about the processes involved in evolution to sustain divisions among experts for another sixty years.[61]

Perhaps Darwin would have waited another fourteen years to go public with his ideas had he not received a manuscript from Alfred Russel Wallace. Fourteen years younger than Darwin and something of a maverick, Wallace had been so impressed by the writings of Humboldt and Darwin that he decided, at age twenty-five, to follow their leads. Earlier Wallace had helped with his brothers' building and civil engineering businesses, taught school, learned enough science to give public lectures, and collected insects as he traveled about the British Isles. English museums and private collectors had created such a demand for insects that both Wallace and his friend in entomology, Henry Bates, thought that they could pay travel expenses by sending home exotic insect specimens. So they set off for the northern reaches of Brazil. It turned out to be hard work in perilous situations—more like Humboldt in the Orinoco basin rather than Darwin aboard the *Beagle*, but the impact of four years of living among Amerindians, Portuguese traders, mestizos, and African slaves proved inestimable to Wallace's intellectual development.

Like many an autodidact, Wallace was easily impressed by big ideas. For him Chambers's *Vestiges* started a long train of thinking about evolution. Chambers gave him the categories for evaluating the biodiversity of the Amazon Valley. The blue macaw was difficult to shoot yet occupied a discrete area. Why, Wallace asked, would there be such a demarcation unless there was something specific in the environment to explain it? Those who would later be called creationists explained features like a bird's beak as fitted to particular food sources, but Wallace found birds with various beaks eating from the same resources. More questions than answers kept him pondering these issues. Returning home proved almost as hazardous as travails along the Amazon. A fire devoured his collection of thousands of bird skins, dried plants, birds' eggs, more than a dozen live birds, and his voluminous notes. Insurance money kept him from total despair and destitution.

During his stay in London Wallace published six papers on the Amazon and met Darwin and Huxley. Within eighteen months he was ready to begin collecting exotic specimens again. This time he chose Malaysia, which was almost virgin territory for European naturalists. What he lacked in education and connections, Wallace made up for in determination. He spent the next six years moving among the islands of the Malay Archipelago. In Borneo he had a chance to observe orangutans in the wild. The influence of *Vestiges* kept the question of their similarity to humans uppermost in his mind.[62] A prolific and popular writer, he published travel literature for the scientifically minded, but Wallace was doing more than reporting: he was puzzling out the origin of the variety of species he encountered in his explorations. By 1858 he was ready to send off to Darwin his manuscript, "On the Tendency of Varieties to Depart Indefinitely From the Original Type."

Wallace's paper exploded a bombshell in Darwin's quiet study. Here was an adumbration of the theory Darwin had been nurturing for two decades. His surprise and chagrin threw him into utter confusion. As they had been gestating in his brain, Darwin had shared his ideas with his old mentor, Lyell, and his close friends Huxley and Joseph Hooker. He had even written the Harvard botanist Asa Gray a long letter in 1857 explaining his theory of natural selection. Wallace had written a manuscript ready for publication that could rob Darwin of his claim to be the author of the ideas that explained the origins of species. Priority was extremely important to scientists then as now. Darwin was shocked by how much it mattered to him, but he also knew that he had spent years working out the arguments that were only hinted at in Wallace's summary.[63] What to do? He consulted his friends. Together, Darwin, Lyell, and Hooker decided that the august Linnean Society should present and then publish both Wallace's manuscript and two unpublished papers by Darwin, including Darwin's letter to Gray, which made an ideal companion to Darwin's 1844 sketch because it incorporated all of Darwin's subsequent ruminations.[64]

That's exactly what they did on the first of July, 1858, one month after Darwin had received Wallace's study. Wallace's work complemented Darwin's evidence of variation in domesticated animals with material on feral ones. They shared the credit for the theory of evolution by natural selection. The Linnean Society members who collected to hear the papers scarcely seemed aware that there was anything special about the evening's program. The society since 1908 has celebrated its role in launching the modern theory of evolution by giving the Darwin-Wallace prize every half century to someone who has made "major advances in evolutionary biology." At the time though, its president concluded his annual report by noting the absence of any "striking discoveries." In later years Wallace turned toward spiritualism and exempted human beings from a purely materialist causality, but he remained a stanch defender of evolution by natural selection and became a treasured friend of Darwin's. By the time of Wallace's death in 1913 he had published three hundred scholarly articles and was considered one of the greatest living scientists.[65]

The next seventeen months were both feverish and decisive for Darwin. The orderly routine at his home didn't change, but the master had. For a book of such importance, the *Origin of Species* had an eccentric, almost haphazard, genesis. Darwin had been refining the core of the argument for over a decade. Now under pressure from Hooker and Lyell, he agreed to forgo his long manuscript and prepare a shortened version of his argument for publication. He called it an abstract; the publisher insisted on dropping the word. Nor did Darwin ever lengthen the work, responding to critics rather than inserting material in subsequent editions. For the remaining twenty-four years of his life Darwin dedicated himself to securing the place of his evolutionary theory.

Haste in this case made for virtuosity. Darwin covered in 490 pages the earth's history, geological patterns of organic change, the geographic and ecological interrelationship of living organisms, along with commentary on instincts, diversity, sexual selection, and inheritance. The *Origin of Species* has an appealing accessibility possibly because of the peculiar way it took shape, rushed into print after a long gestation period. Like Humboldt, Darwin's style was personal, appealing directly to the reader's curiosity. Who could not be charmed by the opening sentence of his masterpiece: "When on board H.M.S. *Beagle*, as naturalist, I was much struck with certain facts in the distribution of the inhabitants of South America, and in the geological relation of the present to the past inhabitants of that continent," or the conclusion, "As this whole volume is one long argument, it may be convenient to the readers to have the leading facts and inferences briefly recapitulated."[66]

When the inevitable outrage came after publication of the *Origin of Species* in November 1859, Wallace fiercely defended it along with Hooker and Huxley. Huxley nicely captured the momentous occasion, when he said that Darwin had addressed "the question of all questions" of how to place humans in nature. All previous mechanisms for evolution had hinged upon metaphysical or teleological properties like vitalistic strivings toward perfection. Even Lamarck had believed in a drive toward improvement, but Darwin and Wallace based their theories exclusively upon natural processes gleaned from empirical evidence. There was something distinctly unorthodox about Wallace, but Darwin represented an incongruous mix of respectability and unrelenting radical impulses.

Darwin's theory involved an intricate dynamic: the proliferation of the young of all species triggering the struggle for the means of existence; the appearance of new traits in individual offspring that improved their adaptability to the environment; the successful passing of these traits through inheritance, allowing for the more likely survival of the animal and its variation; and over time varieties turning into species. The cascading modification of species that Darwin documented accounted for the superabundance of species that had so troubled Linnaeus. The modification of species also explained extinction. The presence of variety now furnished proof of the constant changes that produced new species.[67]

Because there were so many discrete parts in Darwin's explanation of evolution, there were lots of places for people to fall off the train of his thought. Philosophers in the tradition of Jean-Jacques Rousseau and Immanuel Kant saw nature as containing a destiny for the human species that would manifest itself over time, time being an agent of education. Kant was long dead when the *Origin of Species* came out, but his influence cast a long shadow. One of its tenets was that nature provided the conditions in which human beings could fulfill their moral potential. Instead Darwin's depiction conjured up Alfred Lord Tennyson's lines about "Nature, red in tooth and claw."[68]

The necessary compatibility of nature with the moral qualities in human beings acted as an unexamined assumption for many. Darwin's old professor, Sedgwick, accused him of breaking the link between man's moral and physical characters. If successful, Sedgwick warned, Darwin's thesis would "sink the human race into a lower grade of degradation than any into which it has fallen since its written records tell us of its history." Other thinkers extracted other divine qualities to oppose evolution.

Louis Agassiz, the Swiss paleontologist, and even Darwin's mentor Lyell, believed that there was no limitation to the power of God, who had created each species separately. Theirs was a static, unchanging natural world.[69] Darwin's ideas not only crashed into Christian orthodoxy, they roughed up the certainties of many scientists who believed that everything in the created universe had an essence. For essentialists the transformation of one species to another was unthinkable. They clung to Buffon's sterility test—that is, that species could only propagate within themselves.

Richard Owen, a comparative anatomist and one of Britain's most distinguished scientists, authored the angriest denunciation of Darwin's book in a widely read review that masked a shrewd, if nasty, defense of his own idea of anatomical archetypes. Natural selection was too crude a mechanism for him. Owen was not without scholarly resources for such a fight, the most significant being his challenge that Darwin could not cite a single fossil sequence that exhibited the features of descent with modification. Darwin defensively pointed to the paucity of evidence in the geological record. Owen's review arrived at Darwin's home when both Hooker and Huxley were visiting; Owen had long been Huxley's nemesis, so the three had a field day jeering at Owen's ad hominem remarks and flamboyant indignation at what he considered Darwin's foolish ideas.

Fifteen years later Othniel Marsh, a Yale-trained paleontologist, filled in the deficit of records. Going west to plumb the vast fossil remains, he braved the hazards of the American frontier to collect horse fossils. By 1874 he was ready to describe the genealogical transformation of a four-toed species to a one-toed one, with links enough to convince the most skeptical.[70] More insidious a criticism was Owen's charge that Darwin said "I think" and "I believe" too often in his text. Yet readers were beguiled by his personal approach, especially as the circle of readership expanded beyond the scientific community to encompass educated men and women. Controversy, as the publisher predicted, generated sales, and the expanding pool of readers clinched Darwin's reputation.

And then there were the clergy and religious naturalists. Theology and science had achieved a mutually enhancing balance in Great Britain when Church of England leaders interpreted Newton's laws of universal gravitation as proof of a God-ordained orderly system exemplified in politics by England's constitutional monarchy.[71] If accepted as a description of how nature operated, Darwin's ideas would have shattered this intellectual conceit. His theory suggested that Providence worked through a very improvident nature. A few liberal Anglican ministers accepted "natural selection" as an element in God's design, though the astronomer John Herschel didn't help in this compromise by labeling Darwin's theory "the law of higgledy-piggledy."[72]

Most scientists had probably already abandoned belief in the inerrant word of the Bible, but the older and more religiously oriented ones were loath to give up their conviction of a spiritual dimension to human life. By the middle of the nineteenth century, many naturalists could accept that inferior life forms became extinct; they balked at the creation of new species without a creator. To all this Darwin replied confidently in subsequent editions of his work that there was no design that might imply a designer, just processes of descent with modifications.

Huxley became known as "Darwin's bulldog," an appropriate image for a man who seemed to find few things more satisfying than a good fight. Effective as a lecturer and writer, he was outspoken in his anticlerical campaign. He didn't like the scientific old guard represented by Owen either. These attitudes pushed him to enter the fray over Origin of Species even though he entertained some doubts about it. Rarely has a controversy had such a climax as that at the annual meeting of the British Association for Advancement of Science in Oxford in June 1860. The program featured a discussion of evolution by modified descent, but the real issue was the intrusion of the church into scientific matters. Apes or angels, bishops or biologists were the choice on offer.

Bishop Samuel Wilberforce spoke for an hour before the crowd, estimated by some as close to a thousand. No scientist, he drew on Owen's arguments, particularly the one stressing the anatomical difference between the brain cavities of humans and animals. Having crafted a secular sermon filled with wit, Wilberforce couldn't resist ending by asking Huxley if he was related on his grandfather's or grandmother's side to an ape. As Huxley rose, he whispered to his neighbor, "The Lord hath delivered him into mine hands." He quickly dispatched Owen's erroneous contentions before answering Wilberforce's supercilious question, saying, "If I would rather have a miserable ape for a grandfather or a man highly endowed by nature and possessed of great means and influence, and yet who employs those faculties for the mere purpose of introducing ridicule into a grave scientific discussion—I unhesitatingly affirm my preference for the ape." The rest was pandemonium—and history.[73]

A more decorous debate along these lines took place in a series of public programs in Boston between Harvard professors Asa Gray and Louis Agassiz, the Swiss paleontologist. Gray had founded American botany and had become an important correspondent of Darwin's. Agassiz had been one of Humboldt's protégés. Although he remained a committed believer in God, Gray shepherded *Origin of Species* through an authorized publication in the United States. For him Darwin's descent with modification offered a way to challenge Agassiz's metaphysical views, which defined species as parts of a divinely created plan. More provocative was Agassiz's belief in polygenesis, with each race created for its specific location on the globe. Like Huxley, Gray was an academic brawler. He took on Agassiz with a zeal that continued in warring articles, the whole recorded in Boston newspapers that Gray clipped for Darwin.[74] And Gray triumphed. It wasn't until the 1920s that some Americans mounted an attack on evolutionary theory.

Copernicus had shifted the earth from the center of the universe; Darwin decentered human beings. They were no longer the beneficiaries of creation in a line from Adam and Eve and Noah. He reduced their moral status to that of mere fortunate adapters. He had also shown how unstable is the balance between religion and science by refusing to halt when his findings cut the bond between God and man. The ensuing years brought to light just how hostile Darwin was to organized Christianity, but he recognized in an advocate like Gray an empiricist who could reconcile his faith with the conviction that natural processes produced species.

Darwin managed to offend a host of people for a variety of reasons. He rejected destiny as vigorously as he did design, which for the sophisticated had become a way to keep God and toss out Christian fundamentalism. Equally upset were those who didn't think that living organisms could be reduced to physical or chemical mechanisms because of the presence of an immaterial, mysterious element many called the soul. Art conveyed this elusive idea better than articles. Think of its elegant expression in Michelangelo's painting of God giving life to Adam in the ceiling of the Vatican's Sistine Chapel.

Darwin also denied progress, the secular religion of the nineteenth century. Just eight years before the publication of *Origin of Species*, Queen Victoria had opened the great London Crystal Palace exposition. Crowds gathered to see its myriad exhibitions extolling the fruits of ingenuity and industry, most of them English. The queen called it "the greatest day of our history, the most beautiful and imposing and touching spectacle ever seen."[75] Although the casual observer might find progress in the movement from the "hairy, tailed quadruped, probably arboreal in its habits" of Darwinian descent to the suited visitor to the Crystal

Palace, Darwin would have none of it. If there was progress, he saw it as circumstantial, not predetermined. "The inhabitants of each successive period in the world's history have beaten their predecessors in the race for life, and are in so far, higher in the scale of nature," Darwin wrote, adding tartly, "and this may account for that vague yet ill-defined sentiment, felt by many paleontologists, that organization on the whole has progressed."[76] To see a destined path would make history teleological, whereas the facts told Darwin there was only variation, struggle, adaptation, and the passing on to progeny of winning traits.

Realizing that more work needed to be done on the evolutionary thesis, Darwin contented himself that "it is a beginning and that is something."[77] He took a pragmatic stance and asked if any new finding was more easily explained by special creation or by evolution. Unsurprisingly, he usually decided that natural selection could best explain it. While he brought out later editions of his great study, Darwin didn't rest with his one work. In fact, the publication of his masterful theory let lose a succession of pathbreaking publications in zoology and botany that he had been nurturing for decades.

The Descent of Man, which appeared in 1871, caused a bigger stir. After a dozen years *Origin of Species* had inspired fresh work that Darwin now drew on to complete his lifetime effort to explain the origins of humans. In *Descent of Man* he had to delve into language, religion, morality, the emotions, self-awareness, reasoning, and the imagination—all the qualities that set human beings apart from their animal relatives. Darwin dealt with innate qualities, not the social learning made possible by the long period of human dependency. Here too appeared the critical importance of sexual selection, the choosing of mates by qualities that thereafter affected the whole population. *Origin of Species* had prepared the way well. The two-volume *Descent* immediately sold well and came out in all the major European languages.[78]

The year 1859 had another memorable benchmark: the death of Alexander Humboldt. On the cusp of ninety, Humboldt had just finished *Cosmos,* the great summation of his life's inquiries brought together for his popular lecture series. Humboldt's funeral was the largest Berlin had ever seen for a private person. His contemporaries remembered that in his youth his fame had been considered second only to Napoleon's.

Darwin's funeral in 1882 in Westminster Abbey was even grander. Burying in the Abbey a religious skeptic as prominent as Darwin disturbed many Church of England prelates, but not burying there the country's most famous scientist since Isaac Newton was worse. The scientific community galvanized by Huxley prevailed, and Darwin was laid to rest not far from Newton, who probably wouldn't have approved of the collateral damage Darwin's theory wreaked on biblical authority. In truth, lots of English people felt the same way, but they honored the man who had so brilliantly demonstrated the possibilities of human reason.[79]

In the United States, the centennial of Alexander Humboldt's birth turned into a major event. Ten years after Humboldt's death, Americans from every part of the country, of literary or scientific bent, embraced Humboldt as their special cultural hero. They anointed him "the second Columbus." On September 14, the exact centennial of Humboldt's birth, the front page of the *New York Times* carried only one word: Humboldt.[80] Agassiz gave a two-hour address that eschewed Humboldt's fields of mineralogy and botany in order to stress his commitment to freedom, independence, and union. Another speaker credited Humboldt with revealing "to the world the secrets of nature and the laws of the universe," insisting that knowledge led to liberation. Humboldt had claimed American topography grander and more sublime than anything in Europe. He described America as "Nature's Nation."[81]

Humboldt, in his flood of publications over a half century, had driven home the point that America was different, not defective. Emerson's generation of Americans in the mid-1800s had little of the cosmopolitan outlook of their grandfathers' cohort, the men who secured independence for the nation. They knew themselves to be provincials, and it bothered them. Only their political and religious virtues could be touted. There were not great works of art, symphonic compositions, statuary, or fine buildings except for the Capitol, which was still under construction. In his *Cosmos*, Humboldt had said that it was America that led to "the appreciation of the grand and the boundless," which made it possible to see the interconnections of all natural phenomena. After Columbus, Europeans began to think in a "wonderfully enlarged" sphere of ideas.

Humboldt had invested the New World with the excitement of new frontiers of the mind, unique in its abundance of fascinating natural phenomena. Two generations of American intellectuals— Ralph Waldo Emerson, Edgar Allan Poe, Henry David Thoreau, his biographer, Washington Irving, Walt Whitman, Francis Parkman, John Muir, and Franz Boaz—had found the inspiration to speak with an authentic American voice from this Prussian aristocrat. At the centennial commemoration in Boston, Emerson called him "one of those wonders of the world . . . who appear from time to time, as if to show us the possibility of the human mind, the force and range of the faculties—a universal man."[82]

The same, of course, could be said of Darwin. Though the men flourished more than three centuries after Columbus's discovery of America, both Humboldt and Darwin had to return to the New World to find the fossils, topography, animals, and plants that would put in perspective what was already known about Europe, Africa, and Asia. Humboldt and Darwin also differed from many of their contemporaries and successors in their generous social sympathies. They were both outspoken critics of slavery and, more important, they saw the evil of invidious comparisons among the races. Humboldt had been furious when his comments about slavery had been excised from the American publication of his travel account. Humboldt had a greater familiarity with the Africans, their descendants, and Amerindians, but Darwin was equally insistent that there was no difference in capabilities among the world's diverse people. This ran counter to the thrust of nineteenth-century scientific opinion about a racial hierarchy with the progressive whites at the head of the evolutionary march. The exceptional resistance of Humboldt and Darwin to these ideas should be celebrated.

More recently critics have cast Humboldt and Darwin as agents of European imperialism. In a sense they were, with their commitment to the superiority of scientific values. These values became implicated in imperialism and racism, but they also laid the foundation for the life sciences that have done so much to improve the quality of life everywhere, with its contribution to life expectancy and to ameliorating a good deal of the world's misery. More important, seeking the truth about nature has nurtured intellectual curiosity throughout the world, enriching millions of lives directly and indirectly.

Over the course of four centuries, studying natural phenomena became an activity defining western modernity while loosening the hold of religious dogma over scientific inquiry. Even if Columbus had never sailed the Atlantic, the investigative spirit would have eventually asserted itself. Still, finding a new continent with its own array of human cultures delivered a monumental shock to Europeans because the phenomena of the New World could not be absorbed into the existing cosmological knowledge of the sixteenth century, provoking people to think anew their assumptions and categories. After centuries in which the learned of the West studied ancient languages in order to unlock the wisdom stored in Christian and classical texts, natural phenomena began to draw Europeans outdoors, up mountains, and across seas to collect specimens and test hypotheses about how nature's processes worked.

Scientific investigators remapped the earth, figuring out its shape and structure while asking provocative questions about the diverse people that shared their planet. The old aristocratic love of adventure expressed itself afresh in grand, state-sponsored expeditions of exploration. More trade, more wealth, and more focused inquiries paved the way for the stellar scientific accomplishments of Humboldt and Darwin. They, in turn, tested and measured concepts about the planet and its inhabitants that demonstrated conclusively that Columbus had found something far more valuable than the Spice Islands he had sought. Retrospectively, the most significant consequence of the age of discovery is the awakening of Europeans' curiosity about the world in which they lived.

Notes

INTRODUCTION

1   As quoted in Thomas Goldstein, *Dawn of Modern Science: From the Arabs to Leonardo Da Vinci* (Boston: Houghton Mifflin, 1980), 57.

2   As quoted in Krzysztof Pomian, *Collectors and Curiosities: Paris and Venice, 1500–1800*, trans. Elizabeth Wiles-Portier (Cambridge, MA: Polity Press, 1990), 58–59.

3   Eric J. Leed, *The Mind of the Traveler: From Gilgamesh to Global Tourism* (New York: Basic Books, 1991), 179; Peter Harrison, "Curiosity, Forbidden Knowledge and the Reformation of Natural Philosophy in Early Modern England," *Isis* 92 (2001): 270–75.

4   Neil Kenny, *The Uses of Curiosity in Early Modern France and Germany* (Oxford: Oxford University Press, 2004), 106; Lorraine Daston and Katharine Park, *Wonders and the Order of Nature, 1150–1750* (New York: Zone Books, 1998), 124–25; Felipe Fernández-Armesto, *Pathfinders: A Global History of Exploration* (New York: W. W. Norton, 2006), 154–57.

5   Frances Wood's *Did Marco Polo Go to China?* (Boulder, CO: Westview Press, 1996) questioned whether the Polos ever got to China. In the lively debate that ensued, the preponderance of contributing scholars concluded that they probably had.

6   As quoted in Laurence Bergreen, *Over the Edge of the World: Magellan's Terrifying Circumnavigation of the Globe* (New York: Morrow, 2003), 14.

7   Boies Penrose, *Travel and Discovery in the Renaissance, 1420–1620* (Cambridge, MA: Harvard University Press, 1952), 268–69.

8   Fernández-Armesto, ed., *The Times Atlas of World Exploration: 3,000 Years of Exploring, Explorers, and Mapmaking* (New York: HarperCollins, 1991), 64.

9   Bergreen, *Over the Edge of the World*, 15.

10 Fernández-Armesto, *Times Atlas*, 47.

11 Pomian, *Collectors and Curiosities*, 59–60.

12 P. J. Marshall and Glyndwr Williams, *The Great Map of Mankind: British Perceptions of the World in the Age of Enlightenment* (London: Dent, 1982), 26; Anthony Pagden, *Fall of Natural Man: The American Indian and the Origins of*

Comparative Ethnology (Cambridge: Cambridge University Press, 1982), 5–6.

13 Pagden, *European Encounters with the New World: From Renaissance to Romanticism* (New Haven: Yale University Press, 1993), 89.

14 Fernández-Armesto, *Pathfinders*, 122–38.

15 Goldstein, *Dawn of Modern Science*, 21–40.

CHAPTER ONE:
THE NEW WORLD FINDS ITS SCRIBES

1 Alfred W. Crosby, Jr., *The Columbian Exchange: Biological and Cultural Consequences of 1492* (Westport, CT: Greenwood Press, 1972), 75–77.

2 As quoted in John Elliott, *The Old World and the New, 1492–1650* (Cambridge: Cambridge University Press, 1970), 20; Felipe Fernández-Armesto, *Pathfinders: A Global History of Exploration* (New York: W. W. Norton, 2006), 166–60.

3 Alberto Salas, *Tres cronistas de Indias: Pedro Martir, Oviedo, Las Casas* (Mexico City: Fondo de Cultura Económica, 1959), 65, 161; Andrée Collard, trans. and ed., *Bartolomé de las Casas' History of the Indies* (New York: Harper & Row, 1971), 38. Oviedo's last name was Fernández, but he is normally referred to as Oviedo.

4 Kathleen Ann Myers, *Fernández de Oviedo's Chronicle of America: A New History for a New World* (Austin: University of Texas Press, 2007), 8–13; *Gonzalo de Oviedo y Valdes, Historia General y Natural de Las Indias*, ed. J. Natalicio Gonzalez (Asuncion del Paraguay: Editorial Guaranía, 1944), I: 6–11; Antonello Gerbi, "The Earliest Accounts of the New World," in Fredi Chiappelli, Michael J. B. Allen, and Robert L. Benson, eds. *First Images of America: The Impact of the New World on the Old* (Los Angeles: University of California, 1976), 40–41.

5 The *Catálogo Real* appeared sixteen years after the king's death in 1504.

6 Gonzalo Fernández de Oviedo, *Sumario de la Natural Historia de las Indias*, Nicholas del Castillo Mathuieu, ed. (Santafé de Bogotá: Instituto Caro y Cuervo: Universidad de Bogotá, 1995), 3; Myers, *Oviedo's Chronicle of America*, 3–7.

7 Crosby, Jr., *Columbian Exchange*, 45–51; Myers, *Oviedo's Chronicle of America*, 37.

8 Figures vary with different scholarly estimates. See Wilbur R. Jacobs, "The Tip of an Iceberg: Pre-Columbian Indian Demography and Some Implications for Revisionism," *William and Mary Quarterly*, 3d ser., 31

(1974): 123–32; Collard, trans. and ed., *Las Casas' History of the Indies*, 461–62; Crosby, Jr., *Columbian Exchange*, 42–51, 75; Charles C. Mann, *1493: Uncovering the New World Columbus Created* (New York: Knopf, 2011), 11.

9 Amir R. Alexander, *Geometric Landscapes: The Voyages of Discovery and the Transformation of Mathematical Practice* (Stanford: Stanford University Press, 2002), 6–7.

10 Anthony Pagden, *Fall of Natural Man: The American Indian and the Origins of Comparative Ethnology* (Cambridge: Cambridge University Press, 1982), 31ff.

11 Jason M. Yaremko "'Gente bárbara': Indigenous Rebellion, Resistance, and Persistence in Colonial Cuba," *Kacike: Journal of Caribbean Amerindian History and Anthropology* (December, 2006), 5–9; Gustavo Gutiérrez, *Las Casas In Search of the Poor of Jesus Christ*, trans. Robert R. Barr (Maryknoll, NY: Orbis Books, 1993), 90–91.

12 Yaremko, "Gente bárbara," 19.

13 Yaremko, "Gente bárbara," 22.

14 Collard, trans. and ed., *Las Casas' History of the Indies*, x–xvii; Mann, *1493*, 350.

15 As quoted in Lewis Hanke, *All Mankind is One: A Study of the Disputation Between Bartolomé de Las Casas and Juan Ginés Sepúlveda in 1550 on the Intellectual and Religious Capacity of the American Indians* (De Kalb: Northern Illinois University Press, 1974), 7; Collard, trans. and ed., *Las Casas' History of the Indies*, 52–54.

16 Antonello Gerbi, *The Dispute of the New World: The History of a Polemic, 1750–1900*, trans. Jeremy Moyle (Pittsburgh: University of Pittsburgh Press, 2010), 7, 40 (original Italian publication, 1955).

17 Myers, *Oviedo's Chronicle of America*, 33–36.

18 Myers, *Oviedo's Chronicle of America*, 18, 99, 102, 140, 33–36; Hanke, *All Mankind is One*, 43.

19 Myers, *Oviedo's Chronicle of America*, 143.

20 Henry Lowood, "The New World and the European Catalog of Nature," in Karen Ordahl Kupperman, ed., *America in European Consciousness, 1493–1750* (Chapel Hill: University of North Carolina Press, 1995), 308–309.

21 Klaus A. Vogel, "Cultural Variety in a Renaissance Perspective," in Bugge and Rubiés, *Shifting Cultures*, 23.

22 Myers, *Oviedo's Chronicle of America*, 64–75, 90–92.

23 Myers, *Oviedo's Chronicle of America*, 64–75. The quotation comes from J. K. Smith, "Nothing Human is Alien to Me," *Religion*, 26 (1996), 304; Gerbi, *Dispute of the New World*, 28.

24 Myers, *Oviedo's Chronicle of America*, 33–36; Salas, *Tres cronistas*, 80–81; Oviedo, *Sumario*, 158.

25 Hanke, *All Mankind is One*, 30.

26 Myers, *Oviedo's Chronicle of America*, 159–61.

27 Myers, *Oviedo's Chronicle of America*, 75–77, 159

28 Myers, *Oviedo's Chronicle of America*, 174–75.

29 Myers, *Oviedo's Chronicle of America*, 21, 126–31. Part 1 of Oviedo's *General and Natural History* had been published in 1535.

30 http://www.san.beck.org/GPJ13-InternationalLaw.html.

31 Gutiérrez, *Las Casas*, 54–55; J. H. Elliott, *Spain and Its World, 1500–1700: Selected Essays* (New Haven: Yale University Press, 1989), 38.

32 David T. Orique, "Journey to the Headwaters: Bartolomé Las Casas in a Comparative Context," *Catholic Historical Review* 95 (January 2009), 4–5.

33 K. W. Swart, "The Black Legend during the Eighty Years War," in *Britain and the Netherlands: Some Political Mythologies*, vol. 5, R. S. Bromley and E. H. Kossmann, eds. (The Hague: M. Nijhoff, 1975), 52–53.

34 *Octavio Remedio*, 1542, as quoted in Gutiérrez, *Las Casas*, 59; see note 23 for the modern research that has calculated that Hispaniola's population was several hundred thousand, not the millions Las Casas believed it to be.

35 Hanke, *All Mankind is One*, 66–69, 146; Pagden, *Fall of Natural Man*, 109–45; Anne M. DeLong, "Disrupting the Discourse of Conquest: The Suppression of Sepúlveda," The Literature of Justification—New Spain—Essays, digital.lib.lehigh.edu/trial/justification/newspain/essay/.

36 Hanke, *All Mankind is One*, 11; Gutiérrez, *Las Casas*, 320–21; Collard, trans. and ed., *Las Casas' History of the* Indies, ix–xi; Myers, *Oviedo's Chronicle of America*, 21, 126–31; Hugh Honour, *The European Vision of America* (Cleveland: Cleveland Museum of Art, 1975), 21.

37 As quoted in Arthur Helps, *The Life of Las Casas: The Apostle of the Indies* (London: Bell and Daldy, 1868), 276.

38 As quoted in Pagden, *European Encounters with the New World: From Renaissance to Romanticism* (New Haven: Yale University Press, 1993), 63.

39 Pagden, *Fall of Natural Man*, 121–29; Myers, *Oviedo's Chronicle of America*, 29–34; Stephen J. Greenblatt, "Learning to Curse: Aspects of Linguistic Colonialism in the Sixteenth Century," in Chiappelli et al., eds., *First Images of America*, 564.

40 George Sanderlin and Gustavo Gutiérrez, eds., *Witness: Writings of Bartolomé De Las Casas* (New York: Orbis Books, 1971), vii; Hanke, *All Mankind is One*, 58–59; Gutiérrez, *Las Casas*, 24. Las Casas's *History of the Indies* was not published until 1875.

41 Elliott, "Renaissance Europe and America," in Chiappelli et al., eds., *First Images of America*, 1: 13; Rubiés, "Travel Writing and Humanistic Culture: A Blunted Impact," in Peter Mancall, ed., *Bringing the World to Early Modern Europe: Travel Accounts and Their Audiences* (Boston: Brill, 2007).

42 Contemporary geneticists hypothesize that losing this capacity was a trade-off for a more robust immune system. I learned about the axolotl and evolution from the writer Gregory Critzer.

43 Hanke, *All Mankind is One*, 95–98; Gutiérrez, *Las Casas*, 11.

44 As quoted in Delong, "Disrupting the Discourse of Conquest," 3, 13.

45 "Nothing Human is Alien to Me," 303–04; Sabine MacCormack, in Kupperman, ed., *America in European Consciousness*, 96.

CHAPTER TWO:

DISCOVERY OF THE OTHER HALF OF THE GLOBE

1 *Webster's Biographical Dictionary*, Philadelphia, 1943; http://pluckywomen .blogspot.com/2010/01/jeanne-bare-botanists-assistant.html.

2 In the early seventeenth century, a Dutch trader named Cape Horn after the city of Hoorn, whose merchants had funded the expedition. The English then turned the name into Cape Horn.

3 George Sanderlin and Gustavo Gutiérrez, eds., *Witness: Writings of Bartolomé Las Casas* (New York: Orbis Books, 1971), 54–55; Laurence Bergreen, *Over the Edge of the World: Magellan's Terrifying Circumnavigation of the Globe* (New York: Morrow, 2003), 74–75. Behaim's globe has survived and can be seen through the Behaim Digital Globe Project in Vienna.

4 Tim Joyner, *Magellan* (Camden, ME: International Marine, 1994), 82; Bergreen, *Over the Edge of the World*, 30–31.

5 Antonio Pigafetta, *The First Voyage Around the World 1519–1522: An Account of Magellan's Expedition*, ed. Theodore J. Cachey, Jr. (Toronto: University of Toronto Press, 2007), 62.

6 Joyner, *Magellan*, 264–65, 237; Bergreen, *Over the Edge of the World*, 128, lix–lxiv.

7 Pigafetta, *First Voyage Around the World*, 66–62; 101–9; Gerbi, "Earliest Accounts of the New World," in Fredi Chiappelli et al, eds., *First Images of America: The Impact of the New World on the Old*, vol. 1 (Los Angeles: University of California Press, 1974), 41–42.

8 Bergreen, *Over the Edge of the World*, 101–4, 109–10.

9 Antonello Gerbi, "The Earliest Accounts of the New World," in Chiappelli et al., eds., *First Images of America*, 39–40; Hugh Honour, *The European Vision of America* (Cleveland: Cleveland Museum of Art, 1975), 3;

David Beers Quinn, "New Geographical Horizons: Literature," in Chiappelli et al., eds., *First Images of America*, II: 636.

10 As quoted in Pigafetta, *First Voyage Around the World,* 9–11.

11 Pigafetta, *First Voyage Around the World,* 57–58. Today on the island of Mactan the Filipinos annually reenact the slaying of the famous navigator in order to celebrate the native chief, Lapulapu, who successfully protected his people from invaders!

12 Pigafetta, *First Voyage Around the World,* 58–59; Bergreen, *Over the Edge of the World,* 41, 291–96.

13 Pigafetta, *First Voyage Around the World,* 9–11, 47, 54, 119–20, 28; Joyner, *Magellan,* 228. See also Pamela Cheek, *Sexual Antipodes: Enlightenment Globalization and the Placing of Sex* (Stanford, Stanford University Press, 2003); Arthur J. Slavin, "The American Principle from More to Locke," in Chiappelli, et al., eds., *First Images of America,* 145.

14 Pigafetta, *First Voyage Around the World,* 28, 36; Bergreen, *Over the Edge of the World,* 98–99.

15 Bergreen, *Over the Edge of the World,* 341–53; Pigafetta, *First Voyage Around the World,* 98–99.

16 Pigafetta, *First Voyage Around the World,* 92–93, 100.

17 Joyner, *Magellan,* 273; John Elliott, *The Old World and the New, 1492–1650* (Cambridge: Cambridge University Press, 1970), 10.

18 Joyner, *Magellan,* 240–41.

19 Joyner, *Magellan,* 346.

20 Joyner, *Magellan,* 239.

21 Joyner, Magellan, 286–87.

22 Joyner, *Magellan,* 349–51.

23 Joyner, *Magellan,* 273, 221–26, 240, 278.

24 Bergreen, *Over the Edge of the World,* 63.

25 Elliott, *Old World and the New,* 40

26 Klaus A. Vogel, "Cultural Variety in a Renaissance Perspective"; Johanne Boemus on "The Manners, Laws and Customs of All People," in Henriette Bugge and Joan Pau Rubiés, eds., *Shifting Cultures: Interaction and Discourse in the Expansion of Europe* (Münster: Lit, 1995), 23.

27 Anthony Grafton, *New Worlds and Ancient Texts: The Power of Tradition and the Shock of Discovery* (Cambridge, MA: Belknap Press of Harvard University Press, 1992), 129; Elliott, *Old World and the New,* 95; Joyce E. Chaplin, *Subject Matter: Technology, the Body and Science on the Anglo-American Frontier, 1500–1676* (Cambridge, MA: Harvard University Press, 2001), 37–40; K. W. Swart, "The Black Legend during the Eighty Years War," in *Brit-*

ain and the Netherlands: Some Political Mythologies, vol. 5, R. S. Bromley and E. H. Kossmann, eds. (The Hague: M. Nijhoff, 1975), 5: 36–57.

28 Peter C. Mancall, ed., *Travel Narratives from the Age of Discovery: An Anthology* (New York, Oxford University Press, 2006), 27.

29 Myron P. Gilmore, "The New World in French and English Historians of the Sixteenth Century," in Chiappelli et al, eds., *First Images of America*, 519–34; Peter Burke, "America and the Rewriting of World History," in Karen Ordahl Kupperman, ed., *America in European Consciousness* (Chapel Hill: University of North Carolina Press, 1995), 37. Rudolf Hirsch, "Printed Reports on the Early Discoveries"; John Elliott, "Renaissance Europe and America: A Blunted Impact?"; Hildegard Binder Johnson, "New Geographical Horizons," in Chiappelli et al., eds., *First Images of America*, 538–39, 14–16, and 619–25.

30 Hans Belting, *Hieronymus Bosch: Garden of Earthly Delights*, 2nd ed. (Munich: Prestel, 2005), 7 (I am indebted to Debora Silverman for this reference); Hoag Levins, "Social History of the Pineapple: Being the Brief and Colorful Story of a Truly American Plant," www.Levins.com/pine apple.html; Jonathan D. Sauer, "Changing Perception and Exploitation of New World Plants in Europe, 1492–1800," in Chiappelli et al, eds., *First Images of America*, 654; Honour, *European Vision of America*, 4.

31 David Beers Quinn, "New Geographical Horizons: Literature," in Chiappelli et al, eds., *First Images of America*, 654; Honour, *European Vision of America*, 3–7.

32 Honour, *European Vision of America*, 19–20.

33 Johnson, "New Geographical Horizons," in Chiappelli et al, eds., *First Images of America*, 619–25.

34 Boies Penrose, *Travel and Discovery in the Renaissance, 1420–1620* (Cambridge, MA: Harvard University Press, 1952), 241–46.

35 Hirsch, "Printed Reports on the Early Discoveries"; Johnson, "New Geographical Horizons: Concepts"; Quinn, "New Geographic Horizons: Literature," in Chiappelli et al, eds., *First Images of America*, 537–48, 619, 646–47.

36 Jonathan D. Spence, *The Memory Palace of Matteo Ricci* (New York: Viking Penguin, 1984); "A Big Map that Shrank the World," *New York Times*, January 20, 2010, C1.

37 Felipe Fernández-Armesto, *Pathfinders: A Global History of Exploration* (New York: W. W. Norton, 2006), 195.

38 Penrose, *Travel and Discovery in the Renaissance*, 261ff; Norman J. W. Thrower, "New Geographical Horizons: Maps," in Chiappelli et al., eds., *First Images of America*, 659–66.

CHAPTER THREE:
PUBLISHERS SPREAD THE WORD OF THE NEW WORLD

1   Joan Pau Rubiés, "Christianity and Civilization in Sixteenth Century Ethnological Discourse," in Henriette Bugge and Joan Pau Rubiés, eds., *Shifting Cultures: Interaction and Discourse in the Expansion of Europe* (Münster: Lit, 1995), 44–45.

2   Elizabeth L. Eisenstein, *The Printing Press as Agent of Change: Communications and Cultural Transformations in Early Modern Europe*, 2 vols. (Cambridge: Cambridge University Press, 1979), 145.

3   Rudolf Hirsch, "Printed Reports on the Early Discoveries and Their Reception," in Fredi Chiappelli et al., eds., *First Images of America: The Impact of the New World on the Old*, vol. 1 (Los Angeles: University of California Press, 1976), 537–48; Hugh Honour, *The European Vision of America* (Cleveland: Cleveland Museum of Art, 1975), 2–3.

4   Eisenstein, *Printing Press as Agent*, 67; Peter Burke, *A Social History of Knowledge from Gutenberg to Diderot* (Cambridge: Blackwell, 2000), 11.

5   Burke, *Social History of Knowledge*, 22.

6   John H. Elliott, "Renaissance Europe and America: A Blunted Impact?" in Chiappelli et. al., eds., *First Images of America*, 13–14.

7   Honour, *European Vision of America*, 5–6.

8   Geoffrey Eatough, ed. and trans., *Selections from Peter Martyr*, in Geoffrey Symcox, ed., *Repertorium Columbianum*, 5 vols. (Brepols, Belgium: Turnout, 1998), 5: 4–5.

9   Eatough, ed. and trans., *Selections from Peter Martyr*, 72.

10 Antonello Gerbi, "The Earliest Accounts of the New World," in Chiappelli et al., eds., *First Images of America,* 40; Eatough, ed. and trans., *Selections from Peter Martyr*, 519–21.

11 Stephen J. Greenblatt, "Learning to Curse: Aspects of Linguistic Colonialism in the Sixteenth Century," in Chiappelli et al., eds., *First Images of America*, 562; Eatough, ed. and trans., *Selections from Peter Martyr*, xi, 27.

12 David Beers Quinn, "New Geographical Horizons: Literature" in Chiappelli et. al., eds., *First Images of America*, 651–52.

13 The Franciscan Giovanni da Pian del Carpine had gone to the Mongol court earlier and written *History of the Mongols*, which did not circulate widely. Peter C. Mancall, ed., *Travel Narratives from the Age of Discovery: An Anthology* (New York, Oxford University Press, 2006), 214.

14 Antonio Pigafetta, *The First Voyage Around the World 1519–1522: An Account of Magellan's Expedition*, ed. Theodore J. Cachey, Jr. (Toronto: University

of Toronto Press, 2007), xl–xli; Benjamin Keen, "The Vision of America in the Writings of Urbain Chauveton," in Chiappelli et al., eds., *First Images of America*, 110; Mancall, ed., *Travel Narratives from the Age of Discovery* (New York: Oxford University Press, 2006), 208; Rubiés, "Travel Writing and Humanistic Culture: A Blunted Impact," in Mancall, ed., *Bringing the World to Early Modern Europe: Travel Accounts and Their Audiences* (Boston: Brill, 2007), 136–37.

15 Rubiés, "Travel Writing and Humanistic Culture," in Mancall, ed., *Bringing the World to Early Modern Europe*, 141 n. 24

16 Hirsch, "Printed Reports on the Early Discoveries," 538–39; Myron P. Gilmore, "New World in French and English Historians," in Chiappelli et al., eds., *First Images of America*, 523; James A. Williamson, *"Richard Hakluyt," in Richard Hakluyt and His Successors*, ed. Edward Lynam, (London: Hakluyt Society, 1946), 18; Honour, *European Vision of America*, 84.

17 John Gimlette, *Wild Coast: Travels on South America's Untamed Edge* (London: Profile Books, 2011); Amir R. Alexander, *Geometric Landscapes: The Voyages of Discovery and the Transformation of Mathematical Practice* (Stanford: Stanford University Press, 2002), 26–40.

18 Eatough, ed. and trans., *Selections from Peter Martyr*, 39.

19 *The Economist*, December 17, 2011, 52–54.

20 Honour, *European Visions of America*, 89; Harold Jantz, "Images of American in the German Renaissance," in Chiappelli et al., eds., *First Images of America*, 1: 95.

21 Russell Jacoby, *Bloodlust: On the Roots of Violence from Cain and Abel to the Present* (New York: Free Press, 2011), 13–16.

22 Stephen Greenblatt, *Marvelous Possessions: The Wonder of the New World* (Chicago: University of Chicago Press, 1991), 146–49.

23 As quoted in Greenblatt, *Marvelous Possessions*, 150.

24 J. M. Cohen, trans. and ed., *Michel de Montaigne: Essays* (Baltimore: Penguin, 1958), 108–17 [originally published 1572–1573]; Arthur J. Slavin, "The American Principle from More to Locke," in Chiappelli et al., eds., *First Images of America*, 148–49.

25 William C. Sturtevant, "First Visual Images of Native America," in Chiappelli et al., eds., *First Images of America*, 419–20.

26 Jantz, "Images of America in the Germany Renaissance," in Chiappelli et al., eds., *First Images of America*, 95–96.

27 K. W. Swart, "The Black Legend during the Eighty Years War," in *Britain and the Netherlands: Some Political Mythologies*, vol. 5, R. S. Bromley and E. H. Kossmann, eds. (The Hague: M. Nijhoff, 1975), 51–54; Anthony

Grafton, *New Worlds, Ancient Texts: The Power of Tradition and the Shock of Discovery* (Cambridge, MA: Belknap Press of Harvard University Press 1992), 129.

28 Grafton, *New Worlds, Ancient Texts*, 129; Elliott, "Renaissance Europe and America," in Chiappelli et al., eds., *First Images of America*, 19.

29 Eatough, ed. and trans., *Selections from Peter Martyr*, 40.

30 Rubiés, "Christianity and Civilization," in Bugge and Rubiés, eds., *Shifting Cultures*, 35–46, 23.

CHAPTER FOUR:
COLLECTORS, MENAGERIES, AND NATURALISTS

1   C. R. Boxer, *Four Centuries of Portuguese Expansion, 1415–1825: A Succinct Survey, 2nd ed.* (Berkeley: University of California Press, 1969), 14.

2   Toby E. Huff, *Intellectual Curiosity and the Scientific Revolution: A Global Perspective* (New York: Cambridge University Press, 2011), 26–27, 44–47.

3   Amir R. Alexander, *Geometric Landscapes: The Voyages of Discovery and the Transformation of Mathematical Practice* (Stanford: Stanford University Press, 2002), 179–85.

4   John Elliott, *The Old World and the New, 1492–1650* (Cambridge: Cambridge University Press, 1970), 30–31; Jorge Cañizares-Esguerra, *How to Write the History of the New World: Histories, Epistemologies, and Identities in the Eighteenth-Century Atlantic World* (Stanford: Stanford University Press, 2001), 70.

5   Charles C. Mann, *1493: Uncovering the New World Columbus Created* (New York: Knopf, 2011), 301, 242.

6   David Beers Quinn, "New Geographical Horizons," in Fredi Chiappelli, et al., eds., *First Images of America: The Impact of the New World on the Old*, 2 vols. (Los Angeles: University of California Press), 2: 646; Jacob Burckhardt, *Civilization of the Renaissance in Italy*, English edition, (New York: Phaidon, 1937), Part 4, 151–52; Hugh Honour, *The European Vision of America*, (Cleveland: Cleveland Museum of Art, 1975), 91; and Krzysztof Pomian, *Collectors and Curiosities: Paris and Venice, 1500–1800*, trans. Elizabeth Wiles-Portier (Cambridge, MA: Polity Press, 1990), 18.

7   Pomian, *Collectors and Curiosities*, 34–35; Marjorie Swann, *Curiosities and Texts: The Culture of Collecting in Early Modern England* (Philadelphia: University of Pennsylvania Press, 2001), 18–27; Paula Findlen, *Possessing Nature: Museums, Collecting, and Scientific Culture in Early Modern Italy* (Berkeley: University of California Press, 1994), 29.

8   Wilma George, "Alive or Dead: Zoological Collections in the Seventeenth Century," and Henry E. Cooman, "Conchology Before Linnaeus," in Oliver Impey and Arthur MacGregor, eds., *The Origin of Museums: The Cabinet of Curiosities in Sixteenth- and Seventeenth-Century Europe* (Oxford: Clarendon Press, 1985), 8, 192.

9   Silvio A. Bedini, *The Pope's Elephant* (Nashville: J. S. Sanders, 1998), 96–97, 122, 158, 184; Honour, *European Vision of America*, 59, 1.

10 Honour, *European Vision of America*, 58.

11 Jonathan D. Sauer, "Changing Perception and Exploitation of New World Plants in Europe, 1492–1800," in Chiappelli et al., eds., *First Images of America*, 819.

12 Cañizares-Esguerra, *How to Write the History of the New World*, 63–64.

13 Christopher James Pastore, "Expanding antiquity: Andrea Navagero and villa culture in the cinquecento Veneto," 2003, http://repository.upenn.edu/dissertations/AA13087447.

14 Giuseppi Olmi, "Science Honour-Metaphor: Italian Cabinets for the Sixteenth and Seventeenth Centuries," in Impey and MacGregor, eds., *The Origins of Museums*, 8–9.

15 Findlen, *Possessing Nature*, 17–21.

16 George, "Alive or Dead," in Impey and MacGregor, eds., *The Origins of Museums*, 186.

17 Findlen, *Possessing Nature*, 22, 24–26, 36–42.

18 Honour, *European Vision of America*, 67–69.

19 Henry Lowood, "The New World and the European Catalog of Nature," in Karen Ordahl Kupperman, ed., *America in European Consciousness, 1493–1750* (Chapel Hill: University of North Carolina, 1995), 303, 312, 316–317.

20 Kenneth Pomeranz and Steven Topik, *The World That Trade Created: Society, Culture, and the World Economy*, 2nd ed. (Armonk, NY: M. E. Sharpe, 2006), 7.

21 Honour, *European Vision of America*, 57–59.

22 Honour, *European Vision of America*, 3–7.

23 Honour, *European Vision of America*, 102–10, 141. The findings were published in William Piso and Georg Marcgrave's *Historianaturalis Brasiliae* in 1648.

24 Honour, *European Vision of America*, 6, 59.

25 Honour, *European Vision of America*, 72. Capitalization modernized.

26 Huff, *Intellectual Curiosity and the Scientific Revolution*, 193–98.

27 Nicholas Dew, "Reading Travels in the Culture of Curiosity," in Peter

Mancall, ed., *Bringing the World to Early Modern Europe: Travel Accounts and Their Audiences* (Boston: Brill, 2007), 39–47.

28 Dew, "Reading Travels," in Mancall, ed., *Bringing the World to Early Modern Europe*, 44.

29 Huff, *Intellectual Curiosity and the Scientific Revolution,* 193–97.

30 Huff, *Intellectual Curiosity and the Scientific Revolution,* 200; Paul de Kruif, *Microbe Hunters* (New York: Harcourt, Brace, 1926), 8–20.

31 Daston, "Baconian Facts, Academic Civility, and the Prehistory of Objectivity," in Allan Megill, ed., *Rethinking Objectivity* (Durham: Duke University Press, 1994), 42–50.

32 Daston and Katharine Park, *Wonders and the Order of Nature, 1150–1750* (New York: Zone Books, 1998), 115–25.

33 "The Refutation of Philosophies," as quoted in Swann, *Curiosities and Texts,* 59; Daston, "Baconian Facts," in Megill, ed., *Rethinking Objectivity,* 44–47.

34 Bacon, *The New Organum or True Direction Concerning the Interpretation of Nature,* 1620, Book 1: chaps. 23, 39–44.

35 Cañizares-Esguerra, *How to Write the History of the New World,* 146, 195.

36 Hans Baron, "The Querrelle of the Ancients and the Moderns as a Problem for Renaissance Scholarship," *Journal of the History of Ideas* 20 (January 1959):7.

37 As quoted in Swann, *The Culture of Collecting in Early Modern England* (Philadelphia: University of Pennsylvania Press, 2001), 59.

38 Gerard L'E. Turner, "The Cabinet of Experimental Philosophy," in Impey and MacGregor, eds., *Origins of Museums,* 215–22.

39 P. J. Marshall and Glyndwr Williams, *The Great Map of the Mankind; British Perceptions of the World in the Age of Enlightenment* (London: Dent, 1982), 48–56.

40 *Webster's Biographical Dictionary.*

41 Swann, *Curiosities and Texts,* 57.

42 Swann, *Curiosities and Texts,* 196; Olmi, "Science Honour-Metaphor," in Impey and MacGregor, eds., *Origins of Museums,* 10.

43 Barbara M. Benedict, *Curiosity: A Cultural History of Early Modern Inquiry* (Chicago: University of Chicago Press, 2001), 40–46.

44 Dale Poulter, "Phosphorus," http://pubs.acs.org/cen/80th/phosphorus.html.

45 Margaret Jacob and Larry Stewart, *Practical Matter: Newton's Science in the Service of Industry and Empire, 1687–1851* (Cambridge, MA: Harvard University Press, 2004), 38–41; Joel Mokyr, *The Gifts of Athena: Historical Origins*

of the Knowledge Economy (Princeton: Princeton University Press, 2002), 44–45.

46 Mann, *1493*, 207–08.

47 Raz Chen-Morris, "The Fall of Icarus and Kepler's Camera Obscura: The Theme of Forbidden Knowledge in the Sixteenth and Seventeenth Centuries Revisited," www.usyd.edu.au/baroquescience/papers/Icarus_ and_Kepler.doc.

48 See Jacob, *Scientific Culture and the Making of the Industrial West* (New York: Oxford University Press, 1997).

49 Roy Porter, "Introduction," in Stephen Pumfrey, Paolo L.Rossi, and Maurice Slawinski, eds., *Science, Culture, and Popular Belief in Renaissance Europe* (New York: Manchester University Press, 1991), 6–8.

50 Benedict, *Curiosity*, 54–56.

51 Chen-Morris, "Fall of Icarus and Kepler's Camera Obscura"; Peter Harrison, "Curiosity, Forbidden Knowledge and the Reformation of Natural Philosophy in Early Modern England," *Isis* 93 (2001): 276.

52 Carlo Ginzburg, "High and Low: The Theme of Forbidden Knowledge in the Sixteenth and Seventeenth Centuries," *Past & Present* 73 (1976): 38.

53 Daston and Park, *Wonders and the Order of Nature*, 122ff; Ginzburg, "High and Low," 41.

54 Daston and Park, *Wonders and the Order of Nature*, 122ff; Ginzburg, "High and Low," 41; *Leviathan*, Part I, chap. 6, n. 5; as quoted in Daston, "Baconian Facts, Academic Civility, and the Prehistory of Objectivity," in Megill, ed., *Rethinking Objectivity*, 57.

CHAPTER FIVE:
AMATEUR NATURALISTS YIELD TO EXPERTS

1 Peter Burke, "America and the Rewriting of World History," in Karen Ordahl Kupperman, ed., *America in European Consciousness, 1493–1750* (Chapel Hill: University of North Carolina Press, 1995), 47.

2 Russell Levine and Chris Evers, "The Slow Death of Spontaneous Generation (1668–1859)," http://www.accessexcellence.org/RC/AB/BC/Spontaneous_Generation.php.

3 Paul de Kruif, *Microbe Hunters* (New York: Harcourt, Brace 1926), 30–35.

4 De Kruif, *Microbe Hunters*, 41–43.

5 Lorraine Daston, "Baconian Facts, Academic Civility, and the Prehistory of Objectivity," in Allan Megill, ed., *Rethinking Objectivity* (Durham: Duke University Press, 1994), 46–50; Mary Louise Pratt, *Imperial Eyes:*

Travel Writing and Transculturation, 2nd ed. (New York: Routledge, 2007), 26–29.

6   As quoted in Mary Terrall, "Frogs on the Mantelpiece: The Practice of Observation in Daily Life," in Daston and Elizabeth Lunbeck, eds., *Histories of Scientific Observation* (Chicago: University of Chicago Press, 2010), 7.

7   Terrall, "Frogs on the Mantelpiece," in Daston and Lunbeck, eds., *Histories of Scientific Observation*.

8   Terrall, "Frogs on the Mantelpiece," in Daston and Lunbeck, eds., *Histories of Scientific Observation*, 10–11.

9   I am indebted to Gregory Critzer for bringing the axolotl to my attention.

10 Jonathan Z. Smith, "Nothing Human is Alien to Me," *Religion* 26 (1996): 300–305.

11 I am indebted to Lisbet Koerner's *Linnaeus: Nature and Nation* (Cambridge, MA: Harvard University Press, 1999), for the material on Linnaeus unless otherwise noted.

12 Smith, "Nothing Human is Alien to Me," 297.

13 As quoted in "Carl Linnaeus" at http://ucmp.berkeley.edu/history/linnaeus.html; David E. Clark and Joseph F. Williamson, eds., *Sunset New Western Garden Book*, 4th ed. (Menlo Park, CA: Sunset, 1979), 509.

14 Koerner, *Linnaeus*, 23–24.

15 Londa Schiebinger, "Why Mammals Are Called Mammals: Gender Politics in Eighteenth-Century Natural History," *American Historical Review* 98 (April 1993): 382–86.

16 Ernst Mayr, *The Growth of Biological Thought, Diversity, Evolution, and Inheritance*, (Cambridge, MA: Belknap Press of Harvard University Press, 1982), 136.

17 Carl Wennerlind, "From Hartlib to Linnaeus: Science, Spirituality, and Political Economy," Paper delivered at the New World of Projects Conference, Huntington Library, June 23–24, 2012, 13.

18 Wennerlind, "From Hartlib to Linnaeus," 15.

19 James D. Drake, *The Nation's Nature: How Continental Presumptions Gave Rise to the United States of America* (Charlottesville: University of Virginia Press, 2011), 57–58.

20 http://ucmp.berkeley.edu/history/ray.html.

21 James Delbourgo, "Slavery in the Cabinet of Curiosities: Hans Sloane's Atlantic World," http://www.britishmuseum.org/pdf/delbourgo%20essay.pdf.

22 Hans Baron, "The Querelle of the Ancients and the Moderns as a

Problem for Renaissance Scholarship," *Journal of the History of Ideas* 20 (January 1959): 7

23 Mayr, *Growth of Biological Thought*, 181–82.

24 Antonello Gerbi, *The Dispute of the New World: The History of a Polemic, 1750–1900*, trans. Jeremy Moyle (Pittsburgh: University of Pittsburgh Press, 1973), 19 (original Italian publication, 1955); Mayr, *Growth of Biological Thought*, 336.

25 Mayr, *Growth of Biological Thought*, 40–41.

26 Gerbi, *Dispute of the New World*, 5–8; Safier, *Measuring the New World* (Chicago: University of Chicago Press, 2008), 70.

27 Drake, *Nation's Nature*, 59.

28 Gerbi, *Dispute of the New World*, 16–17, 7–8, 20–23.

29 Hugh Honour, *The European Vision of America* (Cleveland: Cleveland Museum of Art, 1975), 234.

30 Gaye Wilson, "Jefferson, Buffon, and the Mighty American Moose," *Monticello Newsletter* 13 (2002); Gerbi, *Dispute of the New World*, 25.

31 Mayr, *Growth of Biological Thought*, 316.

32 Smith, "Nothing Human is Alien to Me," 297.

CHAPTER SIX:
THE TRUE SHAPE OF THE EARTH

1 As quoted in Jonathan D. Sauer, "Changing Perception and Exploitation of New World Plants in Europe, 1492–1800," in Fredi Chiappelli et al., eds., *First Images of America*, 654.

2 Neil Safier, *Measuring the New World: Enlightenment Science and South America* (Chicago: University of Chicago Press, 2008), 6; Charles C. Mann, *1493: Uncovering the New World Columbus Created* (New York: Knopf, 2011), 15.

3 Mary Terrall, *The Man Who Flattened the Earth: Maupertuis and the Science of the Enlightenment* (Chicago: University of Chicago Press, 2002), 92–94; Safier, *Measuring the New World*, 7.

4 Terrall, *Man Who Flattened the Earth*, 7–9; Safier, *Measuring the New World*, 27–28.

5 Terrall, *Man Who Flattened the Earth*, 58.

6 Robert Whitaker, *The Mapmaker's Wife: A True Tale of Love, Murder, and Survival in the Amazon* (New York: Basic Books, 2004), 50.

7 Mary Louise Pratt, *Imperial Eyes: Travel Writing and Transculturation,* 2nd ed. (New York: Routledge, 1992), 16–17; Safier, *Measuring the New World*, 50, 234; Terrall, *Man Who Flattened the Earth*, 92–94, 101–2.

8 Whitaker, *Mapmaker's Wife*, 101–3

9 Terrall, *Man Who Flattened the Earth*, 107–23.

10 Terrall, *Man Who Flattened the Earth*, 126–41.

11 Whitaker, *Mapmaker's Wife*, 101–2.

12 Whitaker, *Mapmaker's Wife*, 162–64.

13 Whitaker, *Mapmaker's Wife*, 287–89, 188–89; Simon Romero, "Once Hidden by Forest, Carvings in Land Attest to Amazon's Lost World," *New York Times*, January 15, 2012, A6. Archaeologists working in the Amazon basin since the 1970s have revised upward their estimates for pre-Columbian population.

14 Terrall, *Man Who Flattened the Earth*, 114–20, 290–91.

15 Whitaker, *Mapmaker's Wife*, 286–87.

16 Whitaker, *Mapmaker's Wife*, 78–70.

17 Gerard L'E. Turner, "The Cabinet of Experimental Philosophy," in Oliver Impey and Arthur MacGregor, eds., *The Origins of Museums: The Cabinet of Curiosities in Sixteenth- and Seventeenth-Century Europe* (Oxford: Clarendon Press, 1985), 215–22.

18 Dava Sobel, *Longitude: The True Story of a Lone Genius Who Solved the Greatest Scientific Problems of His Time*, New York: Walker, 1994, 3–16, 37–39, 51–52.

19 Sobel, *Longitude*, 8–10, 16, 25, 28, 88–89. The quotation appears on p. 8.

20 Terrall, *Man Who Flattened the Earth*, 85.

21 Richard Yeo, *Encyclopaedic Visions: Scientific Dictionaries and Enlightenment Culture* (Cambridge: Cambridge University Press, 2001), 72–75.

22 Safier, *Measuring the New World*, 242–43.

23 Yeo, *Encyclopaedic Visions*, 142–43.

24 Ernst Mayr, *The Growth of Biological Thought, Diversity, Evolution, and Inheritance* (Cambridge, MA: Belknap Press of the Harvard University Press, 1982), 337–38.

25 Jacques Revel, "The Uses of Comparison: Religions in the Early Eighteenth Century," in Lynn Hunt, Margaret Jacob, and Wijnand Mijnhardt, eds., *Bernard Picart and the First Global Vision of Religion* (Los Angeles: Getty, 2010), 341–43.

26 Hunt, Jacob, and Mijnhardt, *The Book that Changed Europe: Picart and Bernard's Religious Ceremonies of the World* (Cambridge, MA: Harvard University Press, 2010), 1–11, 100–3, 296–310.

27 Frank Palmeri, "Conjectural History and Satire: Narrative as Historical Argument from Mandeville to Malthus (and Foucault)," *Narrative* 14 (January 2006): 71.

28 Safier, *Measuring the New World*, 260; Anthony Pagden, *Fall of Natural Man:*

The American Indian and the Origins of Comparative Ethnology (Cambridge: Cambridge University Press 1982), 9–13.

29 P. J. Marshall and Glyndwr Williams, *The Great Map of Mankind: British Perceptions of the World in the Age of Enlightenment* (London: Dent & Sons, 1982), 1.

30 Terrall, *Man Who Flattened the Earth*, 4–5; Robert C. Allen, *The British Industrial Revolution in Global Perspective* (New York: Cambridge University Press, 2009), 10; Joel Mokyr, *The Gifts of Athena: Historical Origins of the Knowledge Economy* (Princeton: Princeton University Press, 2002), 68.

31 Frank Palmeri, "Conjectural History and Satire: Narrative as Historical Argument from Madeville to Malthus (and Foucault)," *Narrative* 14 (2006): 64–65; http://homepace.newschool.edu/het//schools/scottish .htm; Marshall and Williams, *Great Map of Mankind*, 1; Safier, *Measuring the New World*, 2; Hans Baron, "The Querrelle of the Ancients and the Moderns," *Journal of the History of Ideas* 20 (January 1959): 13, 7.

CHAPTER SEVEN:
ATTENTION TURNS TO THE PACIFIC

1 National Mining Association, "The History of Gold," http://www.nma .org/pdf/gold/gold_history.pdf.

2 Bernard Smith, *European Vision and the South Pacific 1768–1850* (New Haven: Yale University Press, 1985), 2.

3 Steven Dutch, "Circumnavigations of the Globe to 1800," www.uwgb .edu/dutchs/westtech/circumn.html.

4 Smith, *European Vision and the South Pacific*, 131; Richard Holmes, *The Age of Wonder: How the Romantic Generation Discovered the Beauty and Terror of Science* (New York: Pantheon, 2009), 3, 33.

5 Andrea Wulf, *Chasing Venus: The Race to Measure the Heavens* (New York: Knopf, 2012); Kenneth Chang, "Look Now for Venus to Cross the Sun, or Wait Another Century," *New York Times*, May 29, 2012, D3.

6 Holmes, *The Age of Wonder*, 5, 17; Rod Edmond, *Representing the South Pacific: Colonial Discourse from Cook to Gauguin* (Cambridge: Cambridge University Press, 1997), 5. Not until the nineteenth century did the people of Tahiti acquire the name "Polynesian," given mainly to distinguish them from the Melanesians farther west.

7 Simon Schaffer, "Visions of Empire: Afterward," in David Philip Miller and Peter Hanns Reill, eds., *Visions of Empire: Voyages, Botany, and Representations of Nature* (Cambridge: Cambridge University Press, 1996), 336–37.

8 Holmes, *Age of Wonder*, 17.

9 Holmes, *Age of Wonder*, 33–34, 37–38.

10 Vanessa Collingridge, *Captain Cook: Obsession and Betrayal in the New World* (London: Ebury Press, 2002), 146–50.

11 David Philip Miller, "Joseph Banks, empire, and 'centers of calculation' in late Hanoverian London," in Miller and Reill, eds., *Visions of Empire*, 31; Holmes, *Age of Wonder*, 57.

12 Collingridge, *Captain Cook*, 127, 216; Christopher Lawrence, "Disciplining disease: scurvy, the navy, and imperial expansion, 1750–1825," in Miller and Reill, eds., *Visions of Empire*, 80.

13 Holmes, *Age of Wonder*, 40–43; Collingridge, *Captain Cook*, 208–11.

14 Alan Bewell, "'On the Banks of the South Sea': Botany and Sexual Controversy in the Late Eighteenth Century," in Miller and Reill, eds., *Visions of Empire*, 176–81; Collingridge, *Captain Cook*, 256, 277.

15 Smith, *European Vision and the South Pacific*, 112, 136–39.

16 Smith, *European Vision and the South Pacific*, 139.

17 Collingridge, *Captain Cook*, 215–19.

18 David Mackay, "Agents of Empire: The Banksian Collectors and Evaluation of New Lands," in Miller and Reill, eds., *Visions of Empire*, 38–53.

19 Daniel O'Sullivan, *In Search of Captain Cook: Exploring the Man Through His Own Words* (London: I. B. Tauris, 2008), 99.

20 O'Sullivan, *In Search of Captain Cook,* 12–13.

21 Smith, *European Vision and the South Pacific*, 18, 56–72, 123; Hugh Cobbe, ed., *Cook's Voyages and People of the Pacific* (London: British Museum, 1979), passim.

22 Collingridge, *Captain Cook*, 328–31.

23 Collingridge, *Captain Cook*, 314–16, 338ff.

24 Smith, *European Vision and the South Pacific*, 118–23; Marshall Sahlins, *Culture and Practical Reason* (Chicago: University of Chicago Press, 1976); Gananath Obeyesekere, *The Apotheosis of Captain Cook: European Mythmaking in the Pacific* (Princeton: Princeton University Press, 1992); Sahlins, *How "Natives" Think about Captain Cook, For Example* (Chicago: University of Chicago Press, 1995); Dorota A. Czarkowska Starzecka, "Hawaii," in Cobbe, ed. *Cook's Voyages and People*, 110.

25 James Burney, *A Chronological History of the Voyages and Discoveries in the South Sea or Pacific Ocean Part II from the year 1579, to the year 1620* (Amsterdam: N. Israel, 1967) [originally published in London, 1806], as cited in Vanessa Smith, "Give Us Our Daily Breadfruit," *Studies in Eighteenth-Century Culture* 35 (2006), n. 2.

26 As quoted in Emma Spary and Paul White, "Food of paradise: Tahitian

breadfruit and the autocritique of European consumption," *Endeavour* 28 (2004): 3; *The Endeavour Journal of Joseph Banks 1768–1771*, J. C. Beaglehole, ed., 2: 341, cited in Smith, "Give Us Our Daily Breadfruit," n. 13; Mackay, "Agents of Empire," in Miller and Reill, eds., *Visions of Empire*, 47–48.

27 Greg Dening, *Mr. Bligh's Bad Language: Passion, Power and Theatre on the Bounty* (Cambridge: Cambridge University Press, 1992), 66.

28 As quoted in Caroline Alexander, *The Bounty: The True Story of the Mutiny on the Bounty* (New York: Penguin, 2003), 124–25.

29 Dening, *Mr. Bligh's Bad Language*, 316–23.

30 Smith, *European Vision and the South Pacific*, 122–23.

31 Edmond, *Representing the South Pacific*, 223–24; Alan Frost, "The Antipodean exchange: European horticulture and imperial designs," in Miller and Reill, eds., *Visions of Empire*, 59–63.

CHAPTER EIGHT:
HUMBOLDT AND DARWIN IN THE NEW WORLD

1 Michael T. Bravo, "Precision and Curiosity in Scientific Travel: James Rennell and the Orientalist Geography of the New Imperial Age (1760–1830)," in Jas Elsner and Joan Pau Rubiés, eds., *Voyages and Visions: Towards a Cultural History of Travel* (London: Reaktion Books, 1999), 162–66.

2 Humboldt's full name was Friedrich Wilhelm Karl Heinrich Alexander von Humboldt.

3 Douglas Botting, *Humboldt and the Cosmos* (New York: Harper & Row, 1973), 113–17.

4 Laura Dassow Walls, *The Passage to Cosmos: Alexander von Humboldt and the Shaping of America* (Chicago: University of Chicago Press, 2009), 27–28.

5 Aaron Sachs, *The Humboldt Current: Nineteenth-Century Exploration and the Roots of America Environmentalism* (New York: Viking, 2006) 53–56.

6 Botting, *Humboldt and the Cosmos*, 28–29.

7 Botting, *Humboldt and the Cosmos*, 53–59; Walls, *Passage to Cosmos*, 35.

8 Alain de Botton, *The Art of Travel* (New York: Vintage 2002), 102.

9 Alexander de Humboldt, *Personal Narrative of Travels to the Equinoctial Regions of the New Continent during the years 1799–1894*, trans. Helen Maria Williams (Philadelphia: M. Carey, 1815), 161–65; Sachs, *Humboldt Current*, 76.

10 Humboldt, *Personal Narrative*, 259–70.

11 Humboldt, *Personal Narrative*, 285–303, 355; Botting, *Humboldt and the Cosmos*, 74–75.

12 Botting, *Humboldt and the Cosmos*, 105ff.

13 F. A. Schwarzenberg, *Alexander von Humboldt, Or What May be Accomplished in a Lifetime* (London, 1866), 105; Humboldt, *Personal Narrative*, 301; Neil Safier, *Measuring the New World: Enlightenment Science and South America* (Chicago: University of Chicago Press, 2008), 60.

14 As quoted in Sachs, *The Humboldt Current*, 2, 12–13, 106–7.

15 Michael Dettelbach, "Global physics and aesthetic empire: Humboldt's physical portrait of the tropics," in David Philip Miller and Peter Hanns Reill, eds., *Visions of Empire: Voyages, Botany and Representations of Nature* (Cambridge: Cambridge University Press, 1996), 258–61; Botton, *Art of Travel*, 103–104; Walls, *Passage to Cosmos*, 127.

16 Botting, *Humboldt and the Cosmos*, 143–46, 96, 147–48.

17 Botting, *Humboldt and the Cosmos*, 149–55; Sachs, *Humboldt's Current*, 355–56; Safier, *Measuring the New World* (Chicago: University of Chicago Press, 2008), xii–xiii.

18 Botton, *Art of Travel*, 114–17; Botting, *Humboldt and the Cosmos*, 155.

19 Botting, *Humboldt and the Cosmos*, 207–9; Sachs, *Humboldt Current*, 52.

20 Charles C. Mann, *1493: Uncovering the New World Columbus Created* (New York: Knopf, 2011), 213–14, 218.

21 Botting, *Humboldt and the Cosmos*, 165; José Enrique Covarrubias, "Political Economy, Alexander von Humboldt, and Mexico's 1810 and 1910 Revolutions," *Rupkatha Journal on Interdisciplinary Studies in Humanities* 2 (2010), 221–22.

22 Botting, *Humboldt and the Cosmos*, 168–72.

23 As quoted in Botting, *Humboldt and the Cosmos*, 175.

24 Botting, *Humboldt and the Cosmos*, 171–75. Spelling and punctuation modernized in these two quotations.

25 Victor Wolfgang von Hagen, *South America Called Them: Explorations of the Great Naturalists La Condamine, Humboldt, Darwin, Spruce* (Boston: Little, Brown & Company, 1955), 162; Botton, *Art of Travel*, 103–10.

26 Covarrubias, "Political Economy, Alexander von Humboldt, and Mexico's 1810 and 1910 Revolutions," 231–32.

27 Walls, *Passage to Cosmos*, 109.

28 Botting, *Humboldt and the Cosmos*, 189.

29 Ian F. McNeely and Lisa Wolverton, *Reinventing Knowledge: From Alexandria to the Internet* (New York: W. W. Norton, 2008), 211–13; Walls, *Passage to Cosmos*, 109; Botting, *Humboldt and the Cosmos*, 221–22, 253; Mary Louise Pratt, *Imperial Eyes: Travel Writing and Transculturation*, 2nd ed. (New York: Routledge, 2008), 118.

30 Botting, *Humboldt and the Cosmos*, 216–17.

31 James S. Aber, "Baron Friedrich W.K.H. Alexander von Humboldt," academic.emporia.edu/aberjame/histgeol/humboldt/humboldt.htm; Walls, *Passage to Cosmos*, 214–16; as quoted in Sachs, *Humboldt Current*, 88.

32 Botting, *Humboldt and the Cosmos*, 277; Walls, *Passage to Cosmos*, ix.

33 Walls, *Passage to Cosmos*, ix.

34 Ernst Mayr, *The Growth of Biological Thought: Diversity, Evolution, and Inheritance* (Cambridge, MA: Belknap Press of Harvard University Press, 1982), 23–24, 121, 442.

35 *Oxford English Dictionary* (Oxford: Oxford University Press, 1971). The new meaning dates to 1802.

36 "Recollections of the Development of my mind and character," in Kenneth Korey, ed., *The Essential Darwin: Selections and Commentary* (Boston: Little, Brown, 1984), 5–10.

37 "Recollections of the Development of my mind and character," in Korey, ed., *Essential Darwin*, 13.

38 "Recollections," in Korey, ed., *Essential Darwin*, 8–13.

39 "Recollections," Korey, ed., *Essential Darwin*, 9.

40 Janet Browne, "Botany in the Boudoir and Garden: the Banksian Context," in Miller and Reill, eds., *Visions of Empire*, 156; Wyhe, ed., Complete Work of Charles Darwin Online, Correspondence, 2002, 3: 140.

41 Botting, *Humboldt and the Cosmos*, 211–13; Pratt, *Imperial Eyes*, 109, 213.

42 Hugh Honour, *The European Vision of America* (Cleveland: Cleveland Museum of Art, 1975), 11–13.

43 Korey, ed., *Essential Darwin*, 18–19.

44 Charles Darwin, *The Voyage of the Beagle* (New York: Dover Publications, 2002), 392–98, 401–402.

45 Walls, *Passage to Cosmos*, 126–27; Peter Hanns Reill, "Seeing and Understanding: A Commentary," in Miller and Reill, eds., *Visions of Empire*, 298–99; Mayr, *Growth of Biological Thought*, 436.

46 "Autobiography," in Korey, ed., *Essential Darwin*, 19.

47 *Journal of Researches into the Natural History and Geology of the Countries Visited During the Voyage of H.M.S. "Beagle" Round the World* (hereafter *The Voyage of the Beagle*); Alison Petch, "Chance and Certitude: Pitt Rivers and his First Collection," *Journal of the History of Collections* 18 (2006): 83.

48 Lisbet Koerner, *Linnaeus: Nature and Nation* (Cambridge, MA: Harvard University Press, 1999), 24, 44–45; L. J. Jordanova, *Lamarck* (Oxford: Oxford University Press, 1984), 2–3, 100–104.

49 Walls, *Passage to Cosmos*, 7, 236; Wyhe, ed., Complete Work of Charles Darwin Online, 9.

50 Korey, ed., *Essential Darwin*, 52, 60, 92, 151–52, 289–303.

51 Mayr, *Growth of Biological Thought*, 339, 347. Paul Eric Olsen, http://rainbow
     .ldeo.columbia.edu/courses/v1001/dinos.2001.html

52 http://ucmp.berkeley.edu/history/ray.html.

53 David Rains Wallace, *Beasts of Eden: Walking Whales, Dawn Horses, and Other
     Enigmas of Mammal Evolution* (Berkeley: University of California Press,
     2005), 8–13. The quotation from Balzac is on p. 13.

54 Botting, *Humboldt and the Cosmos*, 37.

55 Mayr, *Growth of Biological Thought*, 188, 182–84, 257, 363.

56 As quoted in Brown, *Darwin*, 431.

57 Arthur MacGregor, "Exhibiting Evolutionism: Darwinism and pseudo-
     Darwinism in museum practice after 1859," *Journal of the History of Collec-
     tions* 21 (2009): 82.

58 E. J. Browne, *Charles Darwin: The Power of Place* (New York: Knopf, 2003),
     11; Carl Zimmer, "Where did all the flowers come from?," *New York Times,*
     September 8, 2009, D1.

59 Korey, ed., *Essential Darwin*, 20.

60 Mayr, *Growth of Biological Thought*, 24, 46, 203–204, 431, 487–88; Korey,
     i, 52; John R. Durant, "Darwin Unbuttoned," *New York Review of Books,*
     April 26, 1988.

61 Charles Darwin, *The Origin of Species by Means of Natural Selection or the Preser-
     vation of Favoured Races in the Struggle for Life* (1859; repr. New York: London:
     Penguin, 1999), 375.

62 Iain McCalman, *Darwin's Armada: Four Voyages and the Battle for the Theory of
     Evolution* (New York: W. W. Norton, 2009), 260–63.

63 Browne, *Charles Darwin*, 14, 235–39.

64 Mayr, *Growth of Biological Thought*, 339, 176; Londa Schiebinger, "Why
     Mammals Are Called Mammals: Gender Politics in Eighteenth-Century
     Natural History," *American Historical Review* 98 (1993): 385.

65 Alfred Russel Wallace, *Darwinism: An Exposition of the Theory of Natural Selec-
     tion, with some of its Applications* (New York: Macmillan, 1889), iii; Mayr,
     The Growth of Biological Thought, 423; Charles Smith, "Responses to Ques-
     tions Frequently Asked about Wallace," http://people.wku.edu/charles
     .smith/Wallace/FAQ.htm.

66 Darwin, *Origin of Species,* 65, 435.

67 Mayr, *Growth of Biological Thought*, 435–36.

68 Gareth Stedman Jones, "Kant, the French Revolution and the Defini-
     tion of the Republic," in Biancamaria Fontana, ed., *The Invention of the
     Modern Republic* (Cambridge: Cambridge University Press, 1994), 166–

67; the line from Canto 56 of Alfred Lord Tennyson's 1850 poem, *In Memoriam A.H.H.*

69 As quoted in Brown, *Darwin*, 94; Korey, ed., *Essential Darwin*, 53, 236, 233; Mayr, *Growth of Biological Thought*, 443.

70 Browne, *Darwin*, 97–99; 209–11; McCalman, *Darwin's Armada*, 305–7; Wallace, *Beasts of Eden*, 42–53

71 Margaret Jacob, *The Newtonians and the English Revolution, 1689–1720* (Ithaca, NY: Cornell University Press, 1976).

72 Mayr, *Growth of Biological Thought*, 485.

73 Browne, *Darwin*, 114–22.

74 Browne, *Darwin*, 132–35.

75 Nicky Levell, "Reproducing India: International Exhibitions and Victorian Tourism," in Michael Hitchcock and Ken Teague, eds., *Souvenirs: The Material Culture of Tourism* (Aldershot, U.K.: Ashgate, 2000), 37.

76 Darwin, *Origin of Species*, 281–82.

77 Darwin, *More Letters of Charles Darwin*, ed. F. Arwin and A. Seward (London: John Murray, 1903), II: 379.

78 F. Arwin and A. Seward, eds., *More Letters of Charles Darwin*, II: 379; Mayr, *Growth of Biological Thought*, 24–27, 424; Browne, *Darwin*, 325–26, 339–50.

79 Walls, *Passage to Cosmos*, 302–23; McCalman, *Darwin's Armada*, 6–8.

80 Sachs, *Humboldt Current*, 10–12.

81 Walls, *Passage to Cosmos*, 12, 22.

82 Walls, *Passage to Cosmos*, 13–17, 256, 265, 153, 279, 291–94, 212–214; http://RWE.org, The Complete Works of Ralph Waldo Emerson, XI: Miscellanies, XXIV: Humboldt.

Illustration Credits

Frontispiece: *Beagle in the Murray Narrow, Tierra del Fuego* (watercolor on paper), Conrad Martens (1801–78) / Down House, Downe, Kent, UK / © English Heritage Photo Library / The Bridgeman Art Library

Page 3: © The British Library Board / 985.h.14 frontispiece

Page 15: © The British Library Board / G.6628.(1), plate IX

Page 45: *The Discovery of the Strait of Magellan* (colored engraving), Oswald Walter Brierly (1817–94) / Private Collection / Index / The Bridgeman Art Library

Page 69: From Jean de Lery, *Histoire d'un voyage fait en la terre du Bresil, dite Amerique*, Geneva: Vignon, 1600, Bancroft Library, Berkeley, CA

Page 91: © The British Library Board / C.38.i.18, 35

Page 95: Public domain

Page 131: Courtesy of Hunt Institute for Botanical Documentation, Carnegie Mellon University, Pittsburgh, PA

Page 159: Pierre-Louis Moreau de Maupertuis, Engraved by J. Daullé, 1741. © The Trustees of the British Museum

Page 183: Courtesy of the Captain Cook Birthplace Museum, Middlesbrough Council

Page 196: © The British Library Board / Add. 23920, f.54

Page 211: iStockphoto

Index

Page numbers in *italics* refer to illustrations.

Cook, Captain James, *183*, 190,
    193–207
  Antarctic Circle crossed by, 203–4
  Australia located by, 197–98, 201,
    207
  circumnavigation by, 203–4, 207
  crew's welfare as concern of, 199,
    208
  first voyage of, *183*, 193–200, *196*
  George III and, 199, 204
  at Hawaii, 205–7
  Maoris encountered by, *196*, 197–98
  navigating skills of, 193, 198
  Newfoundland coast mapped by,
    193, 199
  at New Zealand, *196*, 197–98
  northwest passage sought by,
    204–5, 206, 207
  second voyage of, 193, 201–4
  at Tahiti, 194–97, 199, 205
  third voyage of, 193, 204–7
  transit of Venus and, *183*, 194–95,
    197
  travel journal of, 199–200
  violent death of, 206–7, 209
Copernicus, Nicolaus, 72, 97, 152, 252
coral reefs, 242
Córdoba, Francisco Hernández de, 16
corn, 17, 108, 109
Corps of Discovery expedition, 224
Cortés, Hernán, 16, 26, 30, 75,
    100–101
  Aztecs brought back to Europe by,
    64, 99
  published letters of, 27–28, 67–68,
    79
  slaves held by, 28
Cosimo II, Grande Duke of Tuscany,
    97
Cosmos (Humboldt), 228–29, 254, 255
Costa Rica, 46
Counter-Reformation, 73
creationism, 244–45
Critzer, Gregory, 263*n*, 272*n*
Crusoe, Robinson (char.), 189–90,
    193
Crystal Palace exposition, 252–53

Cuba, 18, 25, 35, 104, 217, 219
Cueva Indians, 33
Cumaná, 217–18
curiosity, 1–14, 17, 20, 42–43, 49,
    100, 112, 113, 127–28, 163, 173,
    181, 190, 191, 256, 257
  of ancient Greeks, 3, 8–12
  Church's prohibition of, 1–5, 8–10,
    13, 14, 18, 96, 98, 176
  of Humboldt, 214, 218, 219–20
  as unbounded, 130
  voyages of discovery motivated by,
    5–8, 10, 11–14
Cuvier, Georges, 155, 198–99,
    238–40, 242
cyanometer, 217
Cyclopaedia (Chambers), 174, 175

Dalrymple, Alexander, 156, 192–93,
    197, 198, 201
Dampier, William, 192–93
Dare, Virginia, 89
Darwin, Charles, 14, 128, 149,
    230–57
  as agent of imperialism, 256
  allies of, 232, 236, 245, 246, 247, 249
  on *Beagle* voyage, 231–36, 240, 247
  at Cambridge, 231–32
  children of, 241, 243
  correspondence of, 241–42
  destined progress denied by, 252–53
  at Galápagos, 234, 235
  generous social sympathies of,
    255–56
  gradual speciation theory of,
    234–35, 237, 240, 241
  on human origins, 253
  Humboldt and, 233, 234, 236–37
  mathematics disliked by, 232
  medical studies of, 230–31
  publications of, 127, 213, 235–36,
    242, 246–54
  readership attracted by, 249–50
  in South America, 212–13, 233–34
  specimen collection of, 234, 241
  Westminster Abbey burial of, 254
  see also evolution, theory of